SOUNDSCAPE ARCHITECTURE

SOUNDSCAPE ARCHITECTURE

KAREN VAN LENGEN AND JIM WELTY

[following pages]
Interpretive spectrogram image made from recordings of Central Park, *Inside Central Park*, 2024. [Jim Welty]

ORO Editions
Publishers of Architecture, Art, and Design
Gordon Goff: Publisher
www.oroeditions.com
info@oroeditions.com

Published by ORO Editions

Author: Karen Van Lengen and Jim Welty
Designer: Kiri Van Lengen-Welty and Pablo Mandel/Circular Studio
Project Manager: Jake Anderson

10 9 8 7 6 5 4 3 2 1 First Edition

ISBN: 978-1-961856-74-5

Prepress and Print work by ORO Editions Inc.
Printed in China

ORO Editions makes a continuous effort to minimize the overall carbon footprint of its publications. As part of this goal, ORO, in association with Global ReLeaf, arranges to plant trees to replace those used in the manufacturing of the paper produced for its books. Global ReLeaf is an international campaign run by American Forests, one of the world's oldest nonprofit conservation organizations. Global ReLeaf is American Forests' education and action program that helps individuals, organizations, agencies, and corporations improve the local and global environment by planting and caring for trees.

for Kiri Van Lengen-Welty

Preface

Why does the sound of space matter?
Why does listening matter?
Why should we care about the aural nature of architecture and landscape architecture?
And if we do care, how can we develop our awareness and appreciation of these qualities as designers and artists?

This book presents a process of sonic exploration of architecture and landscape architecture, with accompanying texts, design projects, and digital artworks, that creatively address the theme of the sonics of space.

Many architects strive to create spaces that have certain atmospheric qualities. Modulation of light is one example. In the design process, we can understand how light penetrates spaces, using both real models and software programs to develop spatial ideas. However, sound analysis can be elusive in the infancy of this same process, as it relies on so many specific interior details that are yet unknown in these early design phases. In addition, the sonics of any space will vary depending on the occupancies and time of day. Yet sound *is* a primary contributor to the atmospheric qualities of any space.

In *The Aesthetics of Atmospheres*, philosopher Gernot Böhme writes about the human perception of atmospheres. He reminds us that the appreciation of an atmosphere is not a singularly visual endeavor, explaining that atmospheres “are not beings like things; they are nothing without a subject feeling them.”[1] His attention to the interactive aspect of *feeling* an atmosphere is relevant to our work and reminds us of the many attributes of listening. An “aesthetic atmosphere” recognizes a more complex perception of the spaces we inhabit. On this, Böhme writes:

> Today atmospheres may be defined briefly as *tuned space*, i.e. a space with a certain mood. From here two more traits of the theory of atmospheres can be advanced: atmospheres are always something spatial, and atmospheres are always something emotional.[2]

Listening is also a social and a political act. When we listen, we begin the first act of engagement and communication with others. In an increasingly polarized world that often promotes virtual relationships over face-to-face engagements, many of us have become isolated and have lost touch with any sense of commonality or shared human experience. Listening is also critical when we consider our local natural habitats, and it can remind us, in a particularly visceral way, of the fragility of our lived environments, increasingly stressed by global climate change. We ask, therefore, how the act of intentional listening can prompt us to engage more fully in the world and with the people in it; and how architecture and urban design can encourage this type of engagement.

As architects and designers, we are typically familiar with the visual and spatial character of space, yet not as attentive to the role that sound plays in that spatial and emotional perception of the atmospheres we make. For example, Grand Central Terminal, in New York City, has an overwhelming oceanic sound in the central hall that gives the majestic space its unique aural identity. Its calm yet energetic aura reminds us that we are passing through one of the great transportation halls of the world, on our way to somewhere. Another example, the Rose Main Reading Room at the New York Public Library, has a completely different aural experience from that of Grand Central Terminal, one in which highly specific sounds punctuate the intended silence of the library setting. As visitors sit down and stand up from the long reading tables, their heavy oak chairs slide across the tile floor, emitting a sound that reverberates throughout the vast hall like the roar of a lion, recalling the stone lions stationed at the library's entrance on Fifth Avenue. These various "sound objects" can be reassuring, as they remind the visitor of the collective experience of working together, while in the context of individual pursuits. Such examples point to the need to more consciously integrate both sound studies as well as listening processes into the design fields, so that we might better understand this elusive yet essential aspect of atmosphere.

In this book we present our *Soundscape Architecture* project and related works. These projects grew out of our interest in buildings and landscapes that contained distinct sonic characteristics, many of which were not planned but nevertheless have a specific identity. We have addressed this challenging design territory by beginning with the act of listening itself. As we recalibrate our attention to the sounds of environments, we not only appreciate each "tuned space," but we can begin to create an internal library of spatial memories associated with their sounds. We present sonic installations, projects, and art works in this book as demonstrations of listening and creatively responding to the sonic atmospheres of our environment.

The book begins with an Introduction by Karen Van Lengen, covering key historic and contemporary examples of sonic architectural spaces and installations as a background to the work presented. The following four chapters, *Presence*, *Communication*, *Discovery*, and *Art*, each portray individual aspects of the art of listening and its transformative process from the ear back to the eye.

In the first chapter, *Presence*, we demonstrate how real-time listening can heighten our sense of being present in the world, attending to each moment as it passes. This presence bends into memories that can enhance specific perceptions of our spatial encounters. Karen Van Lengen's essay, "Lacrimae Rerum, Tears of Memory: Sonic Encounters in the Public Realm," describes that power of listening in three art installations that deploy real-time recordings and sound devices to commemorate both historical and contemporary events that we cannot see, but whose sounds bring poignant revelations that reach beyond the purely visual. In this chapter we also introduce the *MIX House*, an experimental domestic residence designed and developed for the Vitra Design Museum's *Open House: Architecture and Technology for Intelligent Living*, a traveling international exhibition (2006–9) and accompanying publication that showcased visionary designs for new domestic architecture. The *MIX House*, designed by Karen Van Lengen, Joel Sanders, and Ben Rubin, offers a new way of reinvigorating a suburban domestic environment through the act of listening. We conclude this chapter with a commentary from Barry Bergdoll, who writes about the *MIX House* project and its legacy within the history of architecture.

The second chapter, *Communication*, underscores the role of listening as a fundamental aspect of communication, highlighting design strategies that can foster opportunities for human engagement. Olga Touloumi writes about Harvard University's Woodberry Poetry Room at the Lamont Library, conceived in 1952 by the architectural duo, Aino and Alvar Aalto. The Aaltos envisioned a unique type of space that prioritized the collective experience of listening over the isolation of the alternative small cubicles populating most institutional music and poetry departments of that era. Over half a century later, Karen Van Lengen, Joel Sanders, and Jim Welty designed and constructed *Sound Lounge* (2009–19) at the University of Virginia School of Architecture, elevating this precedent to a new level of engagement in a public institutional setting. Their distinctive group of "sound rooms," defined not by physical walls but by sound itself, stimulated a flow of casual dialogue and exchange to challenge the prevailing "age of isolation" that we encounter in the 21st century.

The third chapter, *Discovery*, is devoted specifically to the *Soundscape Architecture* project, documenting the discoveries born out of the act of listening to iconic architectural interiors

1-1 Digital sound drawing, Guggenheim Museum, from *Soundscape New York*, 2013. [Jim Welty]

and landscapes. The first iteration of this project, *Soundscape New York*, was developed as an installation for the Museum of the City of New York (2015). In this chapter we include author Tony Hiss's opening remarks, presented for this exhibition. We then document our specific process of listening, recording, composing, and drawing those sound compositions. Following this phase, we create animations combining the sound compositions with the visual representations, and finally we generate the digital paintings that recall this entire process. We note many of the discoveries that emerged from this method, and we demonstrate, with analytical diagrams, how architectural characteristics and occupancies shape the sonics of each space. In addition to *Soundscape New York*, we also present two landscape projects, *Inside Central Park* (New York City) and *Openings* (Academical Village at the University of Virginia). In these exterior unbounded spaces, we searched for dominant and unique sounds that characterized the essence of each landscape. We end this chapter with Peter Waldman's creative meditation on this collection of projects.

In the final chapter, *Art*, Nana Last writes about the transformative process of listening into art. We present Jim Welty's digital paintings originally created for an exhibition at the Branch Museum of Architecture and Design in Richmond, Virginia (2020). These final visual compositions reflect our long journeys of being in these spaces and knowing them beyond their normative visual associations. They depict the fleeting and ephemeral aspects of space, what we experience but do not see. Here, Welty recalls these visitations as a poetic engagement with the atmospheric qualities of the spaces as well as with the sonic characters that danced through them on the days of the recordings.

Listening is a beginning—into a fascinating world of information, emotion, and of deep connections to the experience of space.

Karen Van Lengen
Charlottesville
February 2025

Introduction

Listening we will not automatically get to a better world, or a better philosophy. Sound does not hold a superior ethical position or reveal a promised land. But it will show us the world in its invisibility: in the unseen movements beneath its visual organization that allow us to see its mechanism, its dynamic and structure, and the investment of its agency, which might well be dark and forbidding. A sonic sensibility reveals the invisible mobility below the surface of a visual world and challenges its certain position, not to show a better place but to reveal what this world is made of, to question its singular actuality and to hear other possibilities that are probable too, but which, for reasons of ideology, power and coincidence do not take equal part in the production of knowledge, reality, value, and truth.[1]

SALOMÉ VOEGELIN

Sonic studies scholar Salomé Voegelin's description of the value of listening presents a possible world beyond vision that exposes the hidden and unpredictable sonics of space. The *Soundscape Architecture* project grew out of this sensorial awareness. It began with our recollections of interior architectural spaces with unique sonic personalities. The more we listened, the more we discovered. We began to record these spaces and collect them, as architects often do by sketching, documenting their travels through visual memories. This essay introduces both historic and contemporary examples of the relation of sound to architectural space within diverse cultural contexts. These examples help to frame the projects presented in this book.

Though we know the experience of architecture *is* a multisensorial one, contemporary culture challenges this recognition. The proliferation of visual software programs, web images, and media platforms drives both architectural training and practice. In addition, ever-emerging forms of communication, such as smart phones, tablets, and headphones, have taken people out of the public realm and into their own private communicative spaces. Such self-imposed isolation is described by the cultural theorist Michael Bull as a negation of public space, in which "we can choose what we want to hear, screening out the world to create a private auditory universe."[2] This situation is consequential for the future of architecture and urban design. How should architects design contemporary spaces when no one seems to be paying attention to them?

Listening to the sonics of space is a beginning step in this inquiry. It places us in the present and offers possibilities for communication and discovery. Yet, sound is ephemeral and fleeting. It marks time and can never be repeated exactly as we once heard it. However, the act of listening can capture these individual moments and give us agency to recall the memorable atmospheres we find. These "tuned spaces," so aptly described by the late philosopher Gernot Böhme, provide emotional connections to places.[3] We commemorate these "tuned spaces" as the songs of architecture. Our work, from listening-into-art, pays homage to this descriptive aspect of atmosphere.

To more clearly understand the relationship between space and sound, it is helpful to recall some key historical examples from a diversity of cultural contexts and time periods. These examples explore both the synthesis of vision and hearing, and the visual dominance that has prevailed during many historical periods. I conclude by presenting some contemporary examples of listening projects, made possible with current technological innovations. These projects celebrate new kinds of listening experiences and new kinds of atmospheres that have inspired our work in the *Soundscape Architecture* project.

In many early oral societies, primary cultural exchange and the preservation of collective memory took place face to face, with the rhythms and melodies of performing bards weaving the extended storylines of epic poems into coherent, communal events for their audiences.[4] These events occurred in spaces that reinforced the act of listening. For example, circular spaces, such as the Citadel at Mycenae (c.1250 BCE) in Greece, effectively bound the voice into a contained framework and reinforced an environment for listening. This is one of many historic examples of listening spaces that were designed to enhance both the singular and collective voices of the culture. While some of these spaces were intentionally designed, many took advantage of natural formations already present in their landscapes. An interesting example of this can be found in the Paleolithic caves at Lascaux and Font-de-Gaume in southwestern France. Steven Waller, an acoustical archeologist, has speculated that a recognition of the specific aural qualities of these caves may have actually preceded the famous drawings that adorn their walls. Caves that exhibit strong echoes and actively reflect percussive sounds tend to have drawings of bulls and bison, while quieter caves exhibit drawings of the feline family.[5] In these cases, sound itself seems to have generated the visual subjects of the spaces.

2-1 Photograph of the Ħal Saflieni Hypogeum, Malta, taken before 1910. [Richard Ellis]

There are other historical examples of underground rooms that may have been intentionally designed to incorporate specific aural qualities for ceremonial rituals. The Ħal Saflieni Hypogeum (c. 4000–2500 BCE) in Malta, a Neolithic necropolis, is the only known European example of a subterranean "labyrinth."[6] This highly developed 500-square-meter complex, believed to be a burial ground constructed over an extended period of time, is comprised of a series of chambers and rooms arranged on three levels. The oracle chamber, located at the center of the below-grade second level, is believed to have functioned as a ceremonial space. This chamber acts as a resonating hall for the voice. When the frequency of 110 Hertz, which is within the range of an average male voice (90–155 Hertz), is sung into the chamber room, the sound of that voice clearly resonates within the space. This resonance occurs because the wavelength of 110 Hertz is a multiple of the precise dimensions of the chamber, therefore reinforcing that one specific tone. The space also contains a niche located at the end of the chamber that, in turn, amplifies the echoes of the voice. This design may have been intentional, to create specific sonic rituals during burial ceremonies.

In 2008, neurologist Ian Cook, along with a team of University of California scientists, investigated the influence of the 110 Hertz frequency on the brain. They found that when subjects listened exclusively to this frequency, their brain activities changed in notable ways. The language processing section of the brain shut down, while the emotive and mood activity centers were activated. Ancient cultures in other regions beyond the Hypogeum may also have been aware of this phenomenon, as observed in other historic structures such as the burial mound at Newgrange, Ireland.[7]

Western cultures were not alone in making connections between architecture and sound. The Maya, for instance, had a clear understanding and preference for sonic interactions with their architecture. The well-known Temple of Kukulkan (750–1200 CE) at Chichén Itzá, Mexico, emits a chirping sound when activated by a person clapping at the base of the 92-tread stairway. Using sonogram analysis, acoustician David Lubman found a striking similarity between the chirped echo effect of the stairs and the chirp of the quetzal bird, which was significant to Maya culture and was thought to have played an important role in larger ceremonial performances. Lubman hypothesized that the zigzag shadow that descends the stairway during the ceremonial period of the spring equinox provided a visual image not unlike a feather floating down the stair. The induced chirping not only gave meaning to the descending shadow but could also be heard at other nearby sites, thus communicating this important spring equinox ceremony to surrounding neighbors.[8]

2-2 Photograph of the Temple of Kukulkan at Chichén Itzá, Yucatan, Mexico. [Ted Van Pelt]

Almost half a globe away, the Greek theater tradition had already matured. The design lineage to perfect these outdoor performance spaces occurred between the sixth and the fourth century BCE and culminated in a series of exemplary theaters that played a significant role in public life. These outdoor theaters were built into hillsides in such a manner that they took advantage of site topography, prevailing winds, and sun orientation. Critical to any performance was the requirement that all speakers, standing at the focal point of the semicircular space, could be heard clearly by all members of the audience. The Ancient Theatre of Epidaurus, for example, dating from the fourth century BCE, could hold 14,000 people, all of whom would have been able to hear the speakers from the

55 rows of seats. Using specific site planning strategies, along with a studied precision of the seat heights and their reflective absorptive behaviors, as well as the design of the *skene* (a backdrop or scenic wall behind which actors could change), the designers created a theater with astonishing auditory clarity for different kinds of vocal performances. This auditory precision reinforced the Greek theater tradition as a forum for learning and entertainment. This performative tradition, best known for the development of the Greek tragedy, was considered an important part of civic education and brought the community together in a series of shared experiences essential to this early democratic culture. Citizens sat together in an egalitarian mode, not distributed by economic class, to share in this community of learning and entertainment.[9]

2-3 Photograph of the Ancient Greek Theatre at Epidaurus.

The Romans, enamored by Greek architecture and theater design, would subsequently borrow this form and alter it for their own needs. Vitruvius, the renowned engineer and architectural writer of the first century, devoted as much attention in *Ten Books on Architecture* to sound, music, and acoustics as he did to the topics of site design and finish materials. This level of attention to acoustical design is unmatched in current architectural writing and theory. Vitruvius's treatises, along with the Romans' sophisticated development of acoustical principles, demonstrate a keen interest in sound and music as central components of an architectural education. His writings on sound and architecture encompassed two modes, a "proportional mode" and an "actual mode." The "proportional mode" relates the spatiovisual experience of width, height, and depth to the tonal proportions of harmonics, which references a normative, mathematical truth in the form of simple numerical ratios.[10] For example, the use of harmonic proportional systems could be used to delineate spatial proportions in architectural designs. In contrast, the "actual mode," based on studied experience and experimentation, describes how certain physical conditions influence the aural conditions of a space.

Neither the Greeks nor the Romans had a developed science of acoustics in the modern sense, where tone is tied to the rate of vibration and propagation related to the elastic behavior of air. However, the early Greeks, living in a preliterate society, did design their architectural spaces with the aim of enhancing oral communication, as did other early societies. These early formal building preferences would change as literate cultures began to emerge. For example, Hellenistic architecture, after the invention of the Greek phonetic alphabet, saw a massive shift in emphasis from the ear to the eye. The Acropolis of Athens, designed during the time of Pericles, is a renowned example of classical Greek architecture, marked by rectilinear plans with attention to symmetry, mathematical precision, and proportional systems such as the golden ratio. Its site planning favored a subtle perspectival organization of the buildings, rather than buildings formed out of the natural topography and landscape. These transformations were designed to accentuate the visual attributes of the architectural complex, rather than the aural. This ascendency, of the visual over the aural characteristics of architecture, would remain important in many successive periods of architectural history.

One later historical period that synthesized the visual and aural senses was the medieval period in Western Europe. The vast interior spaces of cathedrals and abbeys were constructed out of stone that allowed for long reverberation times. This meant that sounds within the vast halls reverberated for up to 11 seconds, greatly enhancing the dominant style of music during this period—namely, Gregorian chant. This music, coupled with the visual awe of these Gothic cathedrals, worked in a synergetic fashion to create a deeply emotive experience. Ranging from Thoronet Abbey in southern France to the Basilica of Santa Croce in Florence, the abbeys and cathedrals of Europe constituted an array of sacred resonant spaces for expressing the gospel of Christianity. The author Robert Lawlor, touched by the visual and acoustical environment of Thoronet Abbey, wrote:

> This enclosure has been constructed upon a precise, almost uncanny acoustical knowledge. Here each sound, even a pin dropped at the end of the nave some 40 meters away, generates a full range of harmonic overtones ... The nave is so sound-sensitive that one becomes aware that every body movement creates an impact on the volume of air in the chamber.[11]

As the Renaissance emerged at the beginning of the 15th century, it marked a clear shift in sensorial awareness, privileging visual over aural experience. New architectural vocabularies and theories developed as members of the Italian architectural Renaissance, Leon Battista Alberti and Andrea Palladio, developed designs for an architecture that grew out of the discovery of linear perspectival painting. Both Alberti and Palladio referenced the Vitruvian "ideal mode," which emphasized the underlying proportions of harmonic musical structures and applied these systems to architectural proportions.[12] Alberti described music and geometry as fundamentally related, writing that, "music is geometry translated into sound. In music the very same harmonies are audible which inform the geometry of the building."[13] Palladio held a similar view although he theorized less in written text on this subject. Like Alberti, Palladio's buildings were designed using the measurements of harmonic proportions in plan, section, and elevation.

However, there is little evidence that Renaissance architects paired visual spatial design with aural considerations for listening, despite their mathematical interest in harmonic proportions. While Gothic architecture had created a harmonic synthesis of music and space, Renaissance architecture often created a sonic tension between the music and the building. One of Palladio's well-known churches, Il Redentore, in Venice, is a masterpiece of architectural procession and proportion, using light and carved formal spaces that unfold as one processes from the entrance to the nave. The development of these proportional systems was central to Palladian architecture, but there is scant evidence that Palladio was aware of the acoustical aspects of the church, as delineated by Vitruvius's "actual mode." Searching to explore this aspect in more depth, architectural historian Deborah Howard has suggested that, in isolated instances, there may have been more discussion about acoustic qualities than originally thought.[14] Her study of the acoustics of Venetian churches suggests that architects were beginning to notate acoustical conditions that might positively influence the sound of musical spaces, but that these ideas were primarily utilized in the designs of the *ospedaletti* (choir halls of Venetian hospitals and charities). The choirs in Venice had become extremely important and consequently needed spaces that provided optimal sonic qualities. As musical instruments developed in the following centuries, so too did the spaces that housed their performances. The separate discipline of acoustics would begin to develop in the late 19th century, and it is now a highly sophisticated area of engineering primarily devoted to the design of concert halls and large, open interior spaces that require specific acoustical conditions.

An important sonic shift emerged during the Industrial Revolution, which triggered many sensorial changes, especially within the sounds of spaces. The movement from a primarily agrarian to a dominant urban landscape, along with the new economy of speed and mechanization, brought significant social, political, and architectural shifts. Beginning in 18th century England, this revolution would introduce new kinds of energy, building

2-4 Photograph of the interior of Thoronet Abbey, France. [Holger Uwe Schmitt: Licensed under CC BY-SA 4.0]

2-5 *La Rivolta* (The Revolt), painting by Luigi Russolo, 1911.

2-6 Photograph of Luigi Russolo with assistant, Ugo Piatti, and their *intonarumori* ("noise instruments").

materials, workspaces, and transportation systems that significantly altered both visual and sonic experience. Architecture needed to accommodate these momentous changes. The most important and influential feature of this new age was the introduction of new building materials, which would support vast increases in building size, leading to new sensorial experiences. Processed steel and the invention of the elevator gave rise to the American skyscraper. Factories were designed to house hundreds of workers using various types of new equipment. The sounds of these spaces were dramatically different from the rural communities from which many of the workers had migrated. In addition, the urban streetscape now included the sounds of trains, trolleys, and early automobiles set within cities that had become denser, and therefore noisier. Great interior halls, as exemplified by the Crystal Palace in London (1851) and the Chicago World's Fair (1893), announced the beginning of a new interior scale that would continue to grow in the 20th century.

This "birth of noise," described by Futurist artist Luigi Russolo, demanded a sensorial adjustment to the increase and changes in the sounds of everyday life. In his 1913 manifesto, *The Art of Noises*, Russolo rejoiced in this new world and in its potential for innovative forms of music. He wrote to the Italian Futurist composer, Francesco Balilla Pratella:

> My dear Pratella, I submit to your Futurist genius these new ideas, and I invite you to discuss them with me. I am not a musician, so I have no acoustic preferences, nor works to defend. I am a Futurist painter who projects onto a profoundly loved art his will to renew everything. This is why, bolder than the bolder professional musician, totally unpreoccupied with my apparent incompetence, knowing that audacity gives prerogative and all possibilities, I have conceived the renovation of music through the "Art of Noises."[15]

Russolo's manifesto suggested a new collaging of sound and visual art. His Futurist paintings demonstrated movement, using form and color, which described this new world in visual terms. He also defined lists of different kinds of noises out of which he and his assistant Ugo Piatti made a new set of noise instruments that he named *intonarumori*, which played "noise music."[16] They imitated sounds from everyday experiences, as well as some natural sounds. None of the instruments survive today, but Russolo's descriptions of these instruments and accompanying noises are relevant to the history of music and the visual arts of the early 20th century, as they predict many of the changes and inventions that would emerge during this period.

In the field of architecture, the 20th century witnessed the creation of a new design vocabulary that also grew out of this new Machine Age. The use of glass and steel was the bread and butter of the International Style as first promoted by the Bauhaus School in Germany, founded by Walter Gropius in 1919. With his like-minded partners, Gropius developed an interdisciplinary art and design program that sought to bring architecture, design, music, and the fine arts into collaboration with one another by creating an innovative curriculum orchestrated to emphasize these interrelationships through the spirit of making. Several of the faculty artists explored the relation between music and painting. They studied new musical structures, such as the emerging 12-tone technique, to guide their abstract paintings, both in terms of formal relationships and the use of color. In 1911, Wassily Kandinsky and Franz Marc formed the Blaue Reiter group, which proposed a relationship between color and abstract forms. Kandinsky, who befriended the composer Arnold Schoenberg in 1911, sought to relate the sounds of the new 12-tone musical structure with his color theories by developing his own palette of colors based on a taxonomy of sounds. In the same year, Kandinsky and Schoenberg gave the *Sound of Color* concert, combining music with visual expression. Paul Klee, another faculty artist at the Bauhaus School, found a kindred spirit in Kandinsky. Both had trained as musicians, and they used their musical knowledge to structure individual methods of developing abstract form. In referring to Klee and Kandinsky, the art historian Hajo Düchting has noted:

> Music took on the role of mentor by demonstrating the temporal processes within a piece of music and also by serving as a metaphor for a new, creative painting that could reveal the temporal aspect with its own unique means.[17]

This inventive interdisciplinary period at the Bauhaus, which explored the relationship between music, color, form, and painting, did not translate directly into the development of the new International Style of architecture that was emerging from the Bauhaus School during the same period. Experimentation with color and architecture was an important feature of the curriculum, but there is little evidence to support any understanding of the sonic repercussions of the International Style. This architectural movement shaped the dream of creating prefabricated building components that could be mass produced and built as a unifying construction system for an emerging global culture. Its organizing structural principle, the steel frame, could support a variety of interior configurations depending on the program of the building. Because of this independent frame structure, interior walls were set free from the overall building structure, allowing flexibility in the interior plans and sections of buildings. The International Style often included typically flat infill panels to create modular bays. The bays were enclosed by a variety of parallel surfaces often made with glass, metal, or concrete. These highly reflective materials became prone to a range of acoustical problems, such as excessive reverberation, unwanted noise, and unusual resonance patterns that accentuated some frequencies and dampened others. The problem was compounded by one of the main tenets of this movement—namely, the elimination of ornament and other surface treatments, which resulted in spaces with no means to absorb sound energy, creating a condition known as "acoustic glare."

2-7 Photograph of Seagram Building, New York City. [Kiri Van Lengen-Welty]

From a sonic perspective, many examples of the International Style of architecture were not ideal, as they consisted of a series of infinitely reflecting, internally mirrored spaces. In addition, as the historian Emily Thompson has noted, these new 20th century building interiors would soon come to be dominated by the emerging technology of air-conditioning systems that climatized these spaces but also created their own white noise, effectively cutting off exterior sonic environments.[18] Ironically, these sealed glass facade buildings, which often revealed impressive views to outside landscapes, simultaneously denied access to their sonic regions, separating vision and sound as isolated senses. One could make the case here that our general level of environmental awareness would be augmented if our habitable spaces were more sonically connected to the natural environments in which they were built. Having the opportunity to listen to the local environment allows for a stronger connection to these surrounding habitats, and hopefully for a more empathetic relationship with other living beings.

It was during this early to mid-20th-century transformative period in architecture and the visual arts that new musical forms would also emerge. Beginning in the 1940s, the French composer Pierre Schaeffer introduced the idea of using recorded sounds as material for musical compositions. These ideas became the foundation of what came to be known as *musique concrète*. On the heels of this new movement, John Cage, a young composer and conceptual artist influenced by both Russolo and Schaeffer, staged a revolutionary concert in Woodstock, New York, in 1952. During this concert, Cage's friend the composer David Tudor, played Cage's composition, entitled *4'33"*, to a live audience. Tudor walked on stage, sat down at the piano, and closed the keyboard cover. He put his watch on the piano cover, and then sat upright and attentive for 4 minutes and 33 seconds with the audience's full attention. At the end of the period, he rose, bowed, and left the stage. This piece demanded that the audience recalibrate their idea of the concert performance. Cage had led them to listen to the unique sounds all around them, and in so doing, he made the ambient environment ripe for compositional purposes. As a student at Vassar College in the 1970s, my music professor also played this piece to his young audience. I remember that day, and exactly where I was sitting in the lecture hall. My attention shifted from the professor to the atmosphere all around me, to the sounds of my colleagues moving about in their wooden seats, to the dropping of pens and the rearranging of feet. It was an awakening moment that opened new possibilities for both musical composition and for the act of listening, so important to the perception of atmospheres.

In revisiting the concept of atmosphere, particularly as it relates to the sonic attributes of space, we pay homage to John Cage and his revelatory ideas about listening, as they informed our *Soundscape Architecture* project. The American artist Robert Irwin has written eloquently about these perceptual atmospheres that we have sought to introduce into our work as well. Irwin, in his book *Being and Circumstance*, writes:

> *Being and Circumstance*, then, constitute the operative frame of reference for an extended (phenomenal) art activity, which becomes a process of reasoning between our mediated culture (being) and our immediate presence (circumstance). *Being* embodies in you the observer, participant, or user, your complete genetic, cultural, and personal histories as "subsidiary" cues bearing on your "focal" attending (experiencing) of your circumstances, again in a "from-to relation." *Circumstance*, of course, encompasses all of the conditions, qualities, and consequences making up the real context of your being *in* the world. There is embedded in any set of circumstances and your being in them the dynamic of a past and future, what was, how it came to be, what it is, and what it may come to be.[19]

Though Irwin refers here to "focal" attention, his work evolved over time to be inclusive of the sonic as well. His description of "*Circumstance*" resonates with our work, as it embodies the themes that we have developed out of synthesizing the aural and the visual attributes of our spatial experiences.

In this final section, I present four exceptionally interesting precedents as a foreground to the ideas and projects we explore in the following chapters of this book. These projects were made possible through the use of innovative technologies, and each, in its own way, demonstrates an emphasis on the experience of listening and how it can alter our consciousness of the world around us.

Presence

In 1995, David Hykes, founder of the *Harmonic Choir*, gave a concert at The Kitchen, an experimental space for the performative arts in New York City. Hykes and his Choir, performers of overtone singing and chanting, used what was then a state-of-the-art technology to transfer the audio of their overtone singing from one interior space to another, and then back again. The Choir performed their repertoire in The Kitchen's recording space. As the Choir sang, their voices were transmitted live to the Cistercian Thoronet Abbey in France, where they reverberated within its resonant interior. There, the singing was simultaneously recorded and transmitted live, back to the auditorium of The Kitchen. The audience heard a performance that had been acoustically shaped by Thoronet Abbey, rather than by the auditorium of The Kitchen—a displacement of sound and space in real time. David Hykes's performance was a revolutionary idea, as it redefined the audience's sensorial experience of the music as well as their understanding of the importance of the architecture that shaped the sound. This concert was experienced in two different places and expanded the listeners' perception of the pieces sung that evening.

Communication

The *Blur Building*, by American architectural studio Diller Scofidio + Renfro, created for the 2002 Swiss Exposition at Lake Neuchâtel, was a powerful installation that celebrated water in all its forms and was described by the designers as an "architecture of fog."[20] Located in the Lake itself, and accessed by ramps, the structure included thousands of high-powered water nozzles that produced a mass of fog and mist to create a unique atmosphere in which sound and vision were dramatically altered. As visitors entered the fog, they could only see a few feet in front of themselves, and their background soundscape included the hum of the water nozzles that had created the fog. However, it was the voices of the visitors that punctuated this overall sonic experience, and these became

2-8 Photograph of *BLUR Building*, Diller Scofidio + Renfro, 2002 Swiss Exposition, Lake Neuchatel, Switzerland. [Diller Scofidio + Renfro]

moments of discovery, an aural collage of voices and whispers moving in and out of auditory range. This was an all-encompassing environment of creative eavesdropping. Cell phones and earbuds were nowhere to be found. The primary aim of this project was for visitors to be fully present in this atmosphere, and to discover others not through vision but through sound and voice.

Discovery

The *Swiss Sound Pavilion*, designed by Swiss architect Peter Zumthor for the 2000 World Expo in Hanover, Germany, introduced the culture of Switzerland to visitors in a novel manner. Constructed of larch and Douglas fir wood components, intended for reuse after the exhibition, Zumthor created a giant, open labyrinth of corridors and small spaces that downplayed the visual experience in favor of the other senses, especially sound, as the name implies. The presentation of Switzerland was transformed from the prosaic image of a cow standing in front of the snow-capped Swiss Alps to one that evoked the sounds, smells, and haptics of this culture. The musical performers distributed throughout the small spaces in the labyrinth created a sonic background that was offset by the voices of visitors whom one could only occasionally see but could clearly hear, as they ventured

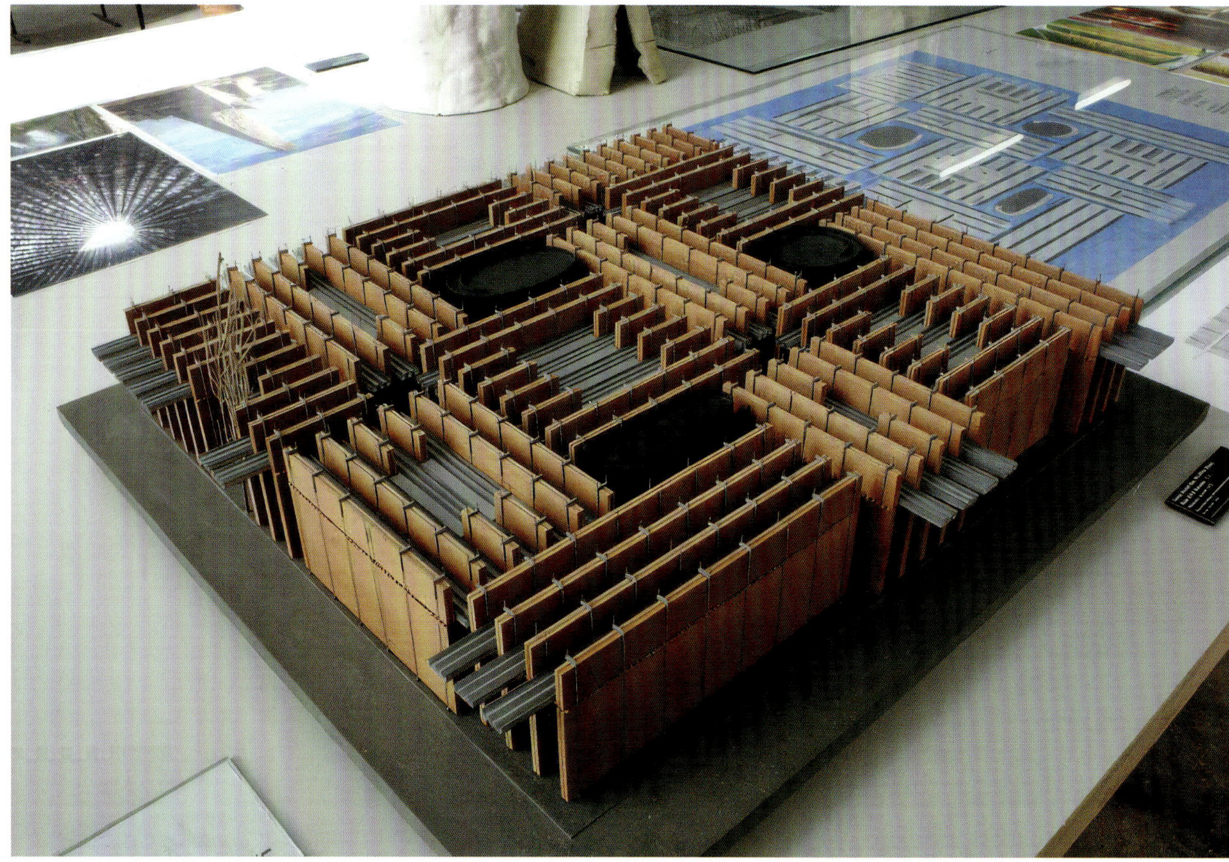

2-9 Photograph of *Swiss Sound Pavilion* model, Peter Zumthor, 2000 World Expo, Hannover, Germany. [Pep Romero Garces]

through this nonhierarchical structure. Again, visitors who fully experienced the sensorial and atmospheric aspects of this exhibition were not engaged in their mobile devices during their visit.

Art *Box with the Sound of Its Own Making*, by American artist Robert Morris, is an artwork created in 1961. Approaching this piece, one first perceives only a pristine wooden box measuring 9¾" × 9¾" × 9¾". Coming closer to the box, one begins to hear the repeating three-hour-long recording that plays from a hidden internal cassette located inside the box. The recording, which incorporates the sounds of the process of making this box, including its sawing, hammering, and sanding, captivate one's attention. This example of the Process art movement, in which artists emphasize the process of making art rather than only the final product, expands our visual and aural perception of the object. It now has a history, a story to convey as to how it came to be, and it tells us that story. In this case, the process is just as important as the final artifact.

These historical and contemporary precedents highlight the significant role of sound in creating memorable atmospheres of spaces and places. In many early examples, we can see that it was necessary to develop architectural forms that could manipulate and enclose sound in order to create communicative systems and significant aural experiences. With the new technologies of the 20th and 21st centuries, those requirements are no longer relevant in the same way. Ironically, it is exactly these new technologies that can often thwart our social, political, and artistic engagement with the world around us. It is not our ambition to ignore these new inventions but instead to deploy them strategically to enhance our experiences of the environments we share. We begin with the act of listening as a first step in this process.

2-10 Photograph of *Box with the Sound of Its Own Making*, Robert Morris, 1961. [Karen Van Lengen]

Presence

But, of course, acoustic space is experienced in real space, as well. It is, however, the space of the body, the space of my own presence, which is expanded through the reach of physical perception. In hearing that tone, voice, and sound do not transfer to the objects from which they may have originated, the listener perceives voice, tone, and sound as a transformation of the space of his own presence. Whoever listens in this manner is dangerously exposed, letting himself out into wide open space, and therefore susceptible to being struck by acoustic events. Sweet melodies may carry him away, claps of thunder may strike him down, buzzing sounds may be threatening to him, a biting sound may injure him. Hearing is a form of being beside oneself and, for this very reason, it can provide a gratifying awareness of being in the world.[1]

GERNOT BÖHME

Lacrimae Rerum, Tears of Memory: Sonic Encounters in the Public Realm

Karen Van Lengen

In the art world, the phrase "sound art" has been coined to describe work that uses sound as a major component of a piece. This term, however, is somewhat problematic as a meaningful category, for it tends to bracket artists and designers who may use sound within their work but who explore very different themes and ideas. Sound, just like visual perception, can act and generate different channels of spatial engagement and provoke a variety of reactions depending on the context and the concept of the piece.

In the 1970s, when the art world was expanding beyond the boundaries of traditional galleries and art museums, public installation art began to emerge and became an area of increasing experimentation for artists and musicians. This kind of artistic endeavor opened up the possibility for provocative site-specific pieces that were intended to intersect with the general public, rather than with an elite museum-going population. Many of these public installations incorporated sound as a primary element to develop the themes of the work. They were also site-specific, transforming the experience of a particular place in the public realm. Salomé Voegelin, an artist and scholar of the sonic arts, has written:

> Sound as a concept invites us into the materiality of things, not to deny the visual but to augment how we might see; and it transgresses the boundaries between the object, the thing looked at, and the space and context of its appreciation, introducing a sense of simultaneity instead of pre-existence, and promoting the reading and experiencing of things as agitational, interventionist, multisensory and capacious.[2]

A sound-generated installation demands that visitors stop and change their sensorial attention to the ear. This engagement with sound is temporal and fleeting and it requires a different kind of concentration. The attention given to something unexpected may be disruptive, but when one gives in to it, it can offer an imaginative flowering of new associations. This individual engagement with the piece or site can also spark a collective reaction as people share their newly discovered experiences with others. The activation of the public realm is a highly valued attribute of these installations, especially in a cultural period in which so much of our public life has turned inward to private networks, headphones, and digital landscapes. If we consider sound to be the agent in this relationship to a site, we can also say that this agent is helping to expose something, to call attention to a particular idea, or to alter our experience of place. In that sonic alteration, there is a change in the atmosphere of the urban condition, which creates an associative memory specific to that location. In many cases, the new memory remains as an active participant in that public space, long after the sound has gone. Sound has the emotive potential to invite us to feel a place, a memory, or an event in a new way. It is an inward experience that is grounded within our bodies and helps us to renegotiate our own personal engagement with a particular place or an event associated with it. This experience opens up the potential for a richer and more nuanced understanding of our public realm and its multiple histories. Kate Lacey, a scholar of media history and theory, refers to this listening as "imaginative visiting," a level of engagement "that involves the imagination in travelling across boundaries physical, personal, social and temporal."[3] This "imaginative" listening can develop into layers of historical memory that are both individual and collective. These are the qualities that give public spaces their character.

This essay examines three different sound installation projects that effectively communicate emotive aspects of historical events and sites through the use of sound, supported by inventive installation strategies. These projects were created to explore significant world events, and have the potential to deeply influence our remembrance and personal engagement with them. This emergent listening can provoke a personal engagement that has the power to temporarily displace our everyday living patterns, allowing us to travel inside of that collective memory. It helps us to more deeply consider important events and allows us to associate those memories with these sites. The three projects presented here are *Sound Island*, by Bill Fontana, *Negative Memorial Aschrottbrunnen*, by Horst Hoheisel, and *Vatnajökull (the sound of)*, by Katie Paterson. They use a diversity of means and methods to achieve an emotive connection to each event, but each project incorporates the sound of water as a primary component of its installation. These projects vary in size and dimension. *Sound Island* was a large, nonpermanent installation that included three different sound presentations all made in real time and located at one of Paris's iconic monuments. *Negative Memorial Aschrottbrunnen* is a permanent installation located in the Courtyard of Honor in Kassel, Germany, while *Vatnajökull (the sound of)*, a limited-time installation, was located in a remote landscape in Iceland but became sonically accessible, through phones, worldwide. As it pertains to each example, I will discuss how the sound of water signifies very different associations within the context of each piece. When we think of the many fountains and water features found in our public parks and urban environments, we often associate the sound of water with peaceful, calming spaces or with playful gathering spots. Yet it can also be a terrifying, violent, and ominous force, and in the following examples the sound of water is the substance of such acoustic experiences; it is the agent of the transformation of memory.

Turbulent Waters

Sound Island, created by Bill Fontana in 1994, commemorated the 50th anniversary of D-Day, when the Allied forces landed on the Normandy beaches to begin the invasion of France during World War II. This commemorative installation was located at the Arc de Triomphe in Paris, one of the city's most important monumental spaces. *Sound Island* was a powerful transformational piece made to honor this pivotal day, one that changed the course of the War and consequently the trajectory of Western European history. On June 6, 1944, the Allies, consisting primarily of British, American, and Canadian troops, staged a surprise landing on five beaches located along the Nazi-occupied Normandy coast of France. More than 155,000 troops arrived by sea and air with the ambition to gain control of northern France. Many of the troops landed in the water, facing heavy gunfire as they moved to take over the beaches. Estimates vary, but it is likely that the Allied casualties on that day totaled over 10,000, while German forces sustained between 4,000 and 9,000 casualties.[4] The successful advance would eventually force the German army out of France and mark the beginning of Germany's retreat and subsequent surrender in 1945. Every year, thousands of tourists visit the Normandy beaches to imagine the day of that invasion, accompanied by stories and memorabilia associated with this famous maneuver.

On the 50th anniversary of D-Day, June 6, 1994, Bill Fontana opened the *Sound Island* installation at the Arc de Triomphe. The Arc itself is layered with history from many different battles and wars. It was initially commissioned by Napoleon to celebrate his military victories following the 1805 Battle of Austerlitz, and in 1920, the Tomb of the Unknown Soldier was added to the space underneath the Arc to honor French soldiers who had died in World War I. It was also from this monument that the Nazis flew their flag in June of 1940 to signify that Paris was under their control. Finally, it was to the Arc that Parisians flocked to celebrate the liberation of Paris in August 1944.[5] Yet Parisians know the Arc as much for its status as a historical landmark as for its urban significance as a primary focal point in the public works program of Georges-Eugène Haussmann, who redesigned the urban infrastructure of 19th-century Paris. Today, the Arc de Triomphe is one of the busiest traffic intersections in the city, its circular base the genesis of 12 streets that radiate outwards to various neighborhoods.

Fontana's project at the Arc was comprised of three different installations: one at street level, one in the tunnel under the street, and one at the top of the monument. Each one provoked a different emotive connection to the events of D-Day. On the street level, Fontana had set up a system that transmitted the sounds of the shores of the Normandy beaches and sent them directly to speakers that washed the walls of the monument with

3-1 Photograph of Parisians lining the Champs Élysées in front of the Arc de Triomphe for a parade on 26 August 1944, the day after the liberation of Paris. [Library of Congress]

these oceanic sounds. From this level, visitors standing under and along the façade of the monument could hear the waves crashing along the Normandy shores located over 150 miles away. This real-time transfer of sound displaced the normal traffic noises at this busy intersection and transported the listeners to another landscape. The sound of beach waves, a sound we often associate with pleasure, was in this case representative of the sonic background for cacophonous gunfire and other sounds of battle. Those waves were the water of terror that day. While the visitor heard only the sound of waves, as they might have sounded minutes before the invasion began, in this act of listening, they could share in the anticipation of what was about to occur on June 6, 1944. Through this kind of imaginative listening, the audience could engage in a shared atmospheric memory of the past, one that connected this particular urban site to the beaches of Normandy and to the angst of so many soldiers who battled their way to the shore that day.

The second part of the installation, located in the tunnel under the street, played underwater sounds from microphones installed on a buoy in the English Channel. Fontana installed these hydrophones in the Channel waters and sent their recordings, again in real time, to the speakers located within the Arc's tunnel space. The sonic quality of the tunnel, as an enclosed arched space, already has a specific acoustical signature, one that has a long reverberation time due to its contained and arched interior shape. The addition of these underwater sounds created an all-encompassing sonic effect. These sounds were reminiscent of what Allied soldiers would have heard that day as they fell into this water of terror, where so many would drown. Walking through this enclosed tunnel had the potential to provoke an overwhelming fright as it transported the listener directly into those waters.

The third part of the installation was located at the top of the Arc de Triomphe, where one can clearly feel the centrality of this monument as the source of those 12 streets radiating out to various Parisian neighborhoods. From a height of 164 feet, or 16 stories, one has an unobstructed view towards these very neighborhoods, which can normally be seen but not heard. On this day, Fontana located microphones in these neighborhoods to be played back to the rooftop speakers. Suddenly the visual connections to these places came alive with the real-time sounds, creating a unique link to these distant views. This kind of urban eavesdropping gave the visitor an intimate connection to the real life of these neighborhoods—to the traffic, the voices, the conversations, the animal sounds, and to other everyday sonic features of urban life. As the artist has proclaimed, "I explored the idea of hearing as far as one can see."[6] This aural connection gave an assurance of the liveliness of these communities that had survived and flourished since the end of the War. Fontana's recordings, 50 years later, sonically confirmed the continuing life force that grew out of the end of the occupation of Paris.

Taken together, all three aspects of this commemorative installation delivered the opportunity for emotive remembrance and sharing, through the precise displacement of site-specific sounds overlaid onto one of Paris's most historic monuments. The real-time aspect is particularly important here because it created a stronger connection to being at the sites of D-Day, affording an immediacy of that experience. It demanded the visitor's presence and encouraged an imaginative transfer to the place and the event. It was not only a great tribute to those who lost their lives that day but offered a different kind of personal connection to European history. The emotive aspects of this installation project did not replace the written commemorative pieces about D-Day, but they augmented one's perception of the events. The sense of "being there," if only for a few moments, altered the personal memory of this historic day and provided opportunities for public engagement in the celebration of its 50th anniversary.

Reflective Waters

Negative Memorial Aschrottbrunnen, designed by Horst Hoheisel, is a permanent Holocaust memorial located in Kassel, Germany. The memorial was commissioned in 1986 and inaugurated the following year. It is unusual among Holocaust memorials as it presents itself as an "anti-memorial," so named by its creator. Most of the Holocaust memorials of the past 70 years have included the names of victims, as well as artifacts and stories intended to bring these events into our consciousness. The Kassel monument is purposely different. Its emptiness first begs us to remember a time in German history when Kassel was an open and vibrant city in which many cultural groups thrived. Located in the Courtyard of Honor at the City Hall of Kassel, the space had once included a fountain, built in 1908 at the behest of a Jewish entrepreneur, Sigmund Aschrott, to celebrate the city and its life within. Children often played in the water, located at the base of the fountain, and their joyful voices gave a sonic liveliness to the courtyard space. The original sandstone obelisk and surrounding fountain, designed by Karl Roth, was destroyed by the Nazis in 1939 because the original patron had been Jewish. In the 1940s, the Nazis then planted the empty stone base with flowers, camouflaging the ruin of the first monument. Later, in 1963, the *Aschrottbrunnen* was rebuilt and the fountain was restored to its former state with a replica of Roth's 1908 design. As many noted, however, this restoration back to the prewar original did not tell the full story of the fountain. The artist Horst Hoheisel, who would go on to design the new monument, advocated to replace the restored fountain with a different kind of memorial that he termed a "negative memorial." Hoheisel has been an important voice in redefining the concept of memorial architecture in Germany, advocating for a more personal engagement with artifacts of memory so that people are obliged to consider past events in the course of their everyday lives. As he recollected:

> During my childhood in Kassel, there were no signs to remind us either of the obelisk designed by Karl Roth or of its founder, Sigmund Aschrott. In Kassel, no one wished to be reminded of the victims of National

> Socialism, of their own guilt, turning to look in the other direction while crimes were being committed. The fountain had become a symbol of memories repressed, the desire to forget.[7]

3-2 Photograph of *Negative Memorial Aschrottbrunnen*, Kassel, Germany. [Simenon]

The memorial that Hoheisel designed is constructed as a sunken fountain, one that has been turned upside down and lodged into the ground. The upside-down obelisk, a replica of the original, has a thin, tall profile with a pointed top, which, when plunged into the ground, brings forth the idea of a deep wound in this urban landscape, a wound of past events and lost lives. What is now visible is only the outline of the original base configuration of the first fountain. Around that base configuration is a narrow, open border that reveals glimpses of the underground space where the fountain now hangs in its inverted position. A visitor walking into the square does not see the fountain, but can see the beginnings of small canals constructed at the edge of the space, transporting water to the site of the monument in the middle of the square. At the edges of this public space, one can also hear the water as it begins its journey to the center of the fountain. Standing directly over the footprint of the fountain's former base, visitors can clearly hear the sound of the churning waters below. It is this turbulent sound of water that causes visitors to stop and listen, to inquire, to find out what lies below the surface. This reaction is made possible by the emptiness of the space and the fullness of the sound of the water, heard but not seen. This is not a quiet water, but water in motion, churning, and seemingly washing the monument below as it rushes around its perimeter. To understand it one must excavate the meaning of the water under this site. The visual emptiness of the site demands that one find out what is below the surface, and only then does the full story of the fountain and its destruction become apparent. Only then is one challenged to listen to the past. At the 1987 opening of the new memorial, Hoheisel commented:

> A simple counter-question is my way of meeting the never-ending stream of attempts to interpret the Shoah. The form that Germans destroyed between 1933 and 1945 can no longer be grasped, either mentally or physically. The destruction of the sandstone form, an "architectural folly" as the architect of City Hall then termed it, was followed by the destruction of the human form. The only way I know to make this loss visible is through a perceptibly empty space, representing the space once occupied. Instead of continuously searching for yet another explanation or interpretation of that which has been lost, I prefer facing the loss as a vanished form. A reflective listening into the void, into the negative of an irretrievable form, where the memory of that which has been lost resounds, is preferable to a mere numb endurance of the facts.[8]

James Young, the founding Director of the Institute for Holocaust, Genocide, and Memory Studies, has written about this project and gives a poignant account of visiting the memorial:

> As we enter the square, we watch as water fills narrow canals at our feet before rushing into a great underground hollow, which grows louder and louder until we finally stand over the "Aschrott Brunnen." Only the sound of gushing water suggests the depth of an otherwise invisible memorial, an inverted palimpsest that demands the visitor's reflection. Through an iron grate and thick glass windows, we peer into the depths.[9]

Reflective listening is the central and key aspect of this project. Without the sounds of water, a visitor would not know that the obelisk lies beneath the surface. A fountain turned upside down is not something that can be seen, but in this memorial the sound of water draws us closer to what lies below. Like the Fontana piece at the Arc de Triomphe, the sound of water here is not soothing, a quality that we often associate with it. Instead, it churns around an artifact of history, as if trying to wash something away that can never be washed away, thus forcing one to confront the memory and loss associated with this historical period. "With the running water," Hoheisel suggests, "our thoughts can be drawn into the depths of history, and there perhaps we will encounter feelings of loss, of a disturbed place, of lost form."[10] The emptiness of this monument in this urban space is powerful and made more so by the sonic quality that is the agent of its memory. Both the *Sound Island* and *Negative Memorial Aschrottbrunnen* projects effectively use sound to create that emotive connection to the past.

Destructive Waters

The last project, *Vatnajökull (the sound of)*, was designed in 2007–8 by Katie Paterson, a Scottish artist whose work addresses geological changes over time and is effective in creating poetic dialogues between nature and culture, between the intimate process of change and the global ramifications of climate change. In 2007 she visited the Jökulsárlón Lagoon, near the Vatnajökull glacier in Iceland, to record the sounds of the melting glacier. She then set up a telephone line from which callers could listen to the melting sounds of the glacier in real time. One could dial the number 07757001122 from anywhere in the world to connect to the recording device. The connection was made possible by the use of a satellite network, and once connected, the caller heard the sound of water dripping, slowly and consistently. After a few minutes of listening to the sound of the glacier melting, callers experienced a feeling of relentlessness and helplessness as they heard the inevitable melting but could not turn the dripping off. Even when the phone connection was no longer there, one imagined that this melting sound would not stop until all the ice was gone. While many people have recorded the sounds of glaciers melting and one can find these recordings on the internet, they do not have the real-time immediacy of this piece, which is what made it so significant. The dripping signified physical loss and the listener experienced that loss directly. Every drop marked a new diminishment of the melting glacier, a process that cannot reverse itself under the current climatic trajectory, caused by human behavior. The piece was a dramatic and powerful reminder of what we are losing. Here again, the water was not a calm agent but instead a relentless force. However, the force was heard and imagined not as an overpowering waterfall, for example, but as one single drip, after another drip, after another drip, and so on. It was the sound of one drip at a time that we know will continue unabated. The piece demanded that we confront the meaning of this situation, not by listening to scientific projections and reviewing data, but by inhabiting the emotional atmosphere that these continuous drips conjure. These drops of water were the lost tears, the crying of a far away landscape.

Unlike the previous examples in Paris and Kassel, this piece was not located in a specific urban setting. A person could access it, and the accompanying sound, from any place in the world using a phone. Imagine, for example, that a person dialed this number while having a waterfront drink in Venice. It would be hard to forget this association after it had been made, because the constant dripping sound symbolizes a global rise in sea level, whose damaging effects will be felt first by Venice and many other coastal cities and landscapes around the world. Every drip became a memory. Every drip was a part of the process of creating global environmental history. This piece, therefore, served the dual purpose of listening to the memory of a glacier that is in the process of melting, and also forecasting an ominous future world that will be created by this melting process.

Conclusion

In thinking about these three examples, one might ask if the use of sound is a more effective generator of remembrance than a visual presentation of the same historical events. Would a person who read about D-Day or saw historical photographs of the invasion be equally moved by its visual documentation? Facts and photographs are always important, as they place these events within their proper context. But sounds, without the visuals, require more of our imagination—of what it might have been like that day. Was it cold or foggy, did the journey through the water seem heavy given the weight of all the equipment and weaponry, did soldiers look straight ahead or look to their comrades for reassurance? In Kassel, we can imagine the sounds of children who once played in the fountain while their parents went about their business in or near City Hall. Could they imagine what would happen here and what their future would hold once this fountain was destroyed? Finally, the dripping sound of the melting glacier brings forth the anguish and immediacy of rising sea levels, and the terrifying possibilities of global climate change that accompany this warning. It is this pure sound, without images, which creates the background for one's imagination and for a deeply personal connection to this ongoing event. Sound can augment what we know by creating that emotional overlay. As one considers all three installations, it is possible to understand how powerful sound can be in the activation of memory and how it can shift our sense of intimacy with history. The development of new technologies to record and transfer sound has driven many of these effective projects and it affords artists the opportunity to create new sound-driven installations. They are the songs of a different kind of information that we experience from inside of ourselves, and they allow us to form new knowledge about places and new personal memories of what we thought we already knew.

3-3 Photograph of Jökulsárlón Glacier Lagoon, Iceland. [Giuseppe Milo]

MIX House

Karen Van Lengen, Joel Sanders, Ben Rubin

In 2006, the Vitra Design Museum in Weil am Rhein, Germany, and the ArtCenter College of Design in Pasadena, California, sponsored the *Open House: Architecture and Technology for Intelligent Living* exhibition to promote new ideas in domestic architecture. Karen Van Lengen, Joel Sanders, and Ben Rubin created the *MIX House* project, an experimental house design that both sees and hears its surrounding landscapes. This design proposal was represented through drawings, an animation, and an installation for a traveling exhibition, beginning in Pasadena, California, and then traveling to Essen, Germany; Warsaw, Poland; and Oslo, Norway. The interior of the installation structure included an animation that fully described the unusual attributes of this house. The *MIX House* animation is now included in the permanent collection of the Art Institute of Chicago.

MIX House explores the possibility of closely coordinating sound and vision, with the goal of enhancing an individual's audiovisual experience of their domestic landscape. This residential dwelling is conceived as a dynamic space enriched by acoustic links to its external environments, which also provide for the integration of new channels of communication within the house.

As noted earlier, Western architecture since the Renaissance has privileged the visual over all other senses, specifically negating the role of the aural environment that had been such a primary aspect of earlier cultures. In recent history, architectural modernism gave rise to the structural frame. This technology allowed designers to exploit the use of expansive glass windows that afforded uninterrupted views of the landscape, while simultaneously applying new acoustical technologies that homogenized these interiors and compromised the aural specificity of both space and place. *MIX House* rejects this privileging of the visual, by putting sound and sight on equal footing. The project proposes a dwelling that rethinks and extends the modernist notion of visual transparency, afforded by the ubiquitous glass window, to include aural transparency as well. The *MIX House* design cohesively

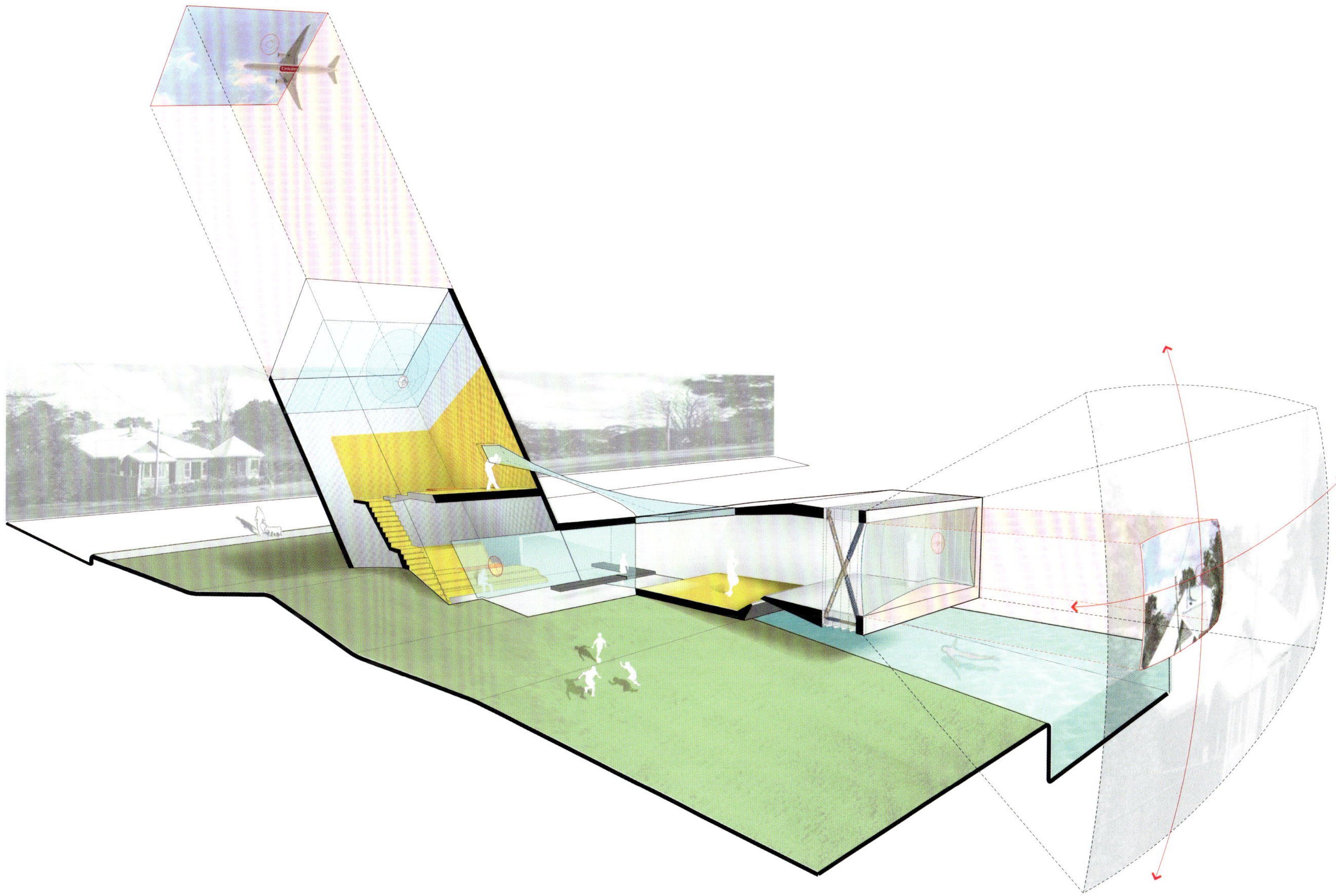

3-4 Axonometric view as seen from the backyard, *MIX House*, 2006. [Joel Sanders, Karen Van Lengen, Ben Rubin]

incorporates cutting-edge technologies and traditional acoustic principles to create a home that constructs and frames audiovisual scenes, enabling occupants to transcend their spatial boundaries.

Situated on a generic suburban plot in Charlottesville, Virginia, the dwelling is composed of two distinct volumes that frame audiovisual scenes adjacent to the house. The backyard façade consists of custom curved operable sonic windows, which capture the outdoor sounds in warm weather. When closed, their curvature can also regulate ambient and specific sounds as well as satellite transmissions. Functioning like an audiovisual telescope, the sonic windows include both a microphone and small video camera located at their centers. The microphone records targeted sounds in real time, and then transmits them to an interior audio system that distributes the sounds to various speakers throughout the house. The video simultaneously records the scene and transmits it to a visual screen located at the kitchen island or sound command center of the home, where the synchronized sights and sounds of the surrounding landscapes may be activated and arranged. The occupants may operate the audiovisual dish to highlight a specific activity or event. They also have the option of putting the system on automatic pilot, permitting them to passively listen to a series of focused sounds captured by the windows and their microphones. In this passive mode, the occupants can design their own rotation and filtering programs, which are able to selectively record specified frequencies such as human voices, birds, or the sound of water. Each sonic window opens to a distinct landscape, affording different opportunities for sound transmissions of these spaces.

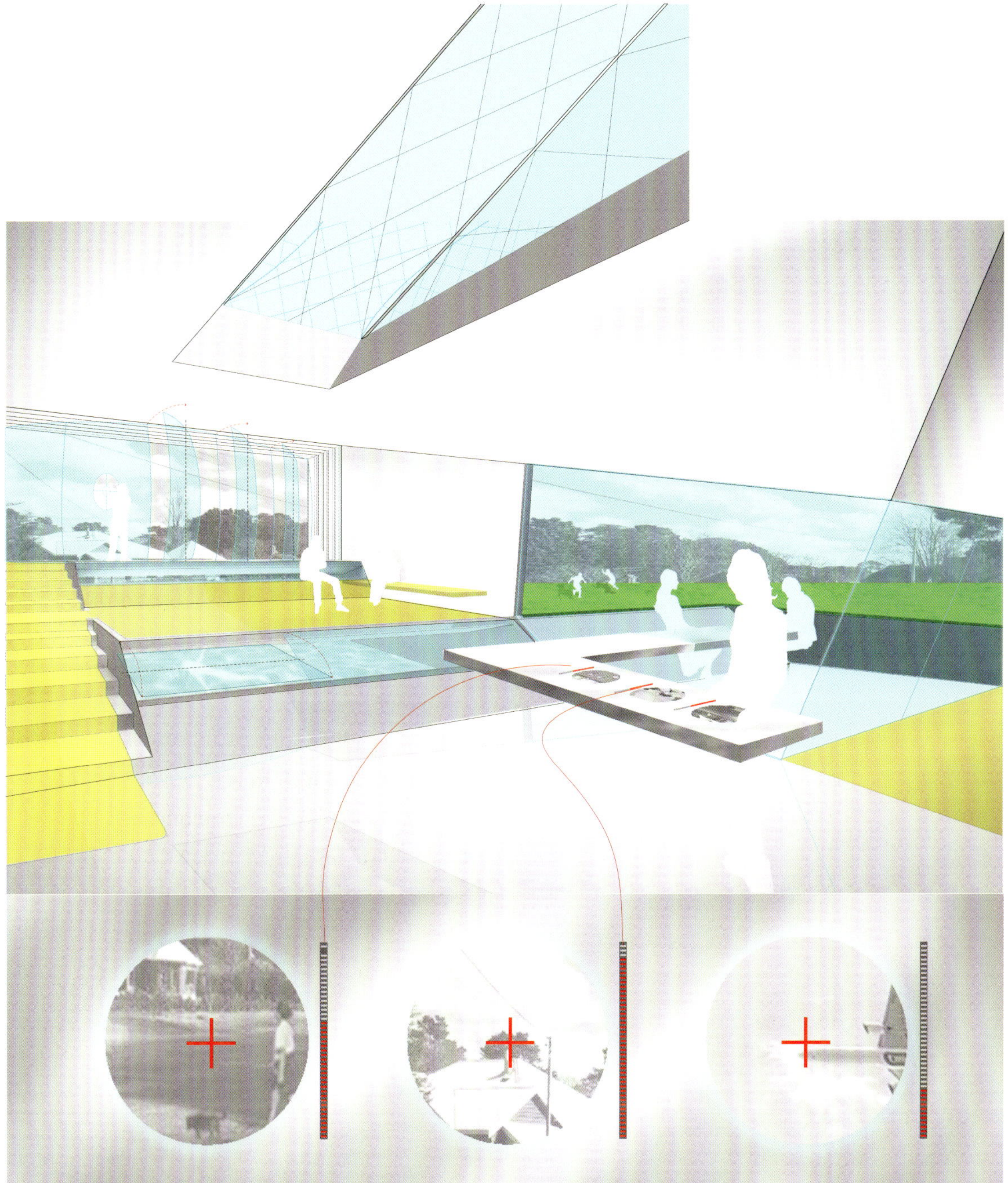

3-5 View of the kitchen area, showing the MIX counter. This is the sound command center of the house, where the synchronized sights and sounds of surrounding landscapes may be recorded and arranged. Operable glass windows under the raised living area can be opened to hear the sounds of the swimming pool below. [Joel Sanders, Karen Van Lengen, Ben Rubin]

The Glass Skylight

This glass skylight located near the top of the vertical volume is designed to both capture or muffle the ambient sounds of the neighborhood. The resulting soft "sonic breeze" is appropriately located just above the sleeping area. In this location, the associated disc doubles as a satellite dish that is embedded in the south-facing section of the skylight. This captures the signals that allow occupants in the bedroom/den below to connect visually and aurally with the global media, through television and internet connections.

The Sonic Picture Window

This sonic window facing the backyard is designed like a camera bellows. It includes a translucent glass dish that is fully integrated into the window wall with the ability to rotate freely in three dimensions. Again, the large windows regulate the ambient sounds while the rotating dish focuses on specific activities of the backyard: the birds at the feeder, the children playing in the sandbox, or the dog chewing on his bone.

The Sonic Front Entry

The curved sliding section of the front sonic window allows residents to hear the ambient sounds of the streetscape. Here, the curved surface doubles as the front door. It can be operated to pick up and highlight the sounds of this locale: guests arriving at the house, the dog greeting the mailman, or the footsteps of the occasional jogger.

The two volumes of the house come together at the center kitchen island, which acts as the audiovisual nerve center of the home and functions as a central command station for operating the sonic windows. From this position the occupants can regulate the various audiovisual conditions associated with the three windows. Each sonic window has a corresponding video screen with regulation controls. The video screen visually records the location of the incoming focused sounds so that the occupants can both see and hear the specific audiovisual condition. This command center, designed into the waterproof kitchen countertop, encourages the occupants to mix these various sounds of their landscapes with other media inputs. For example, the voices of their children playing

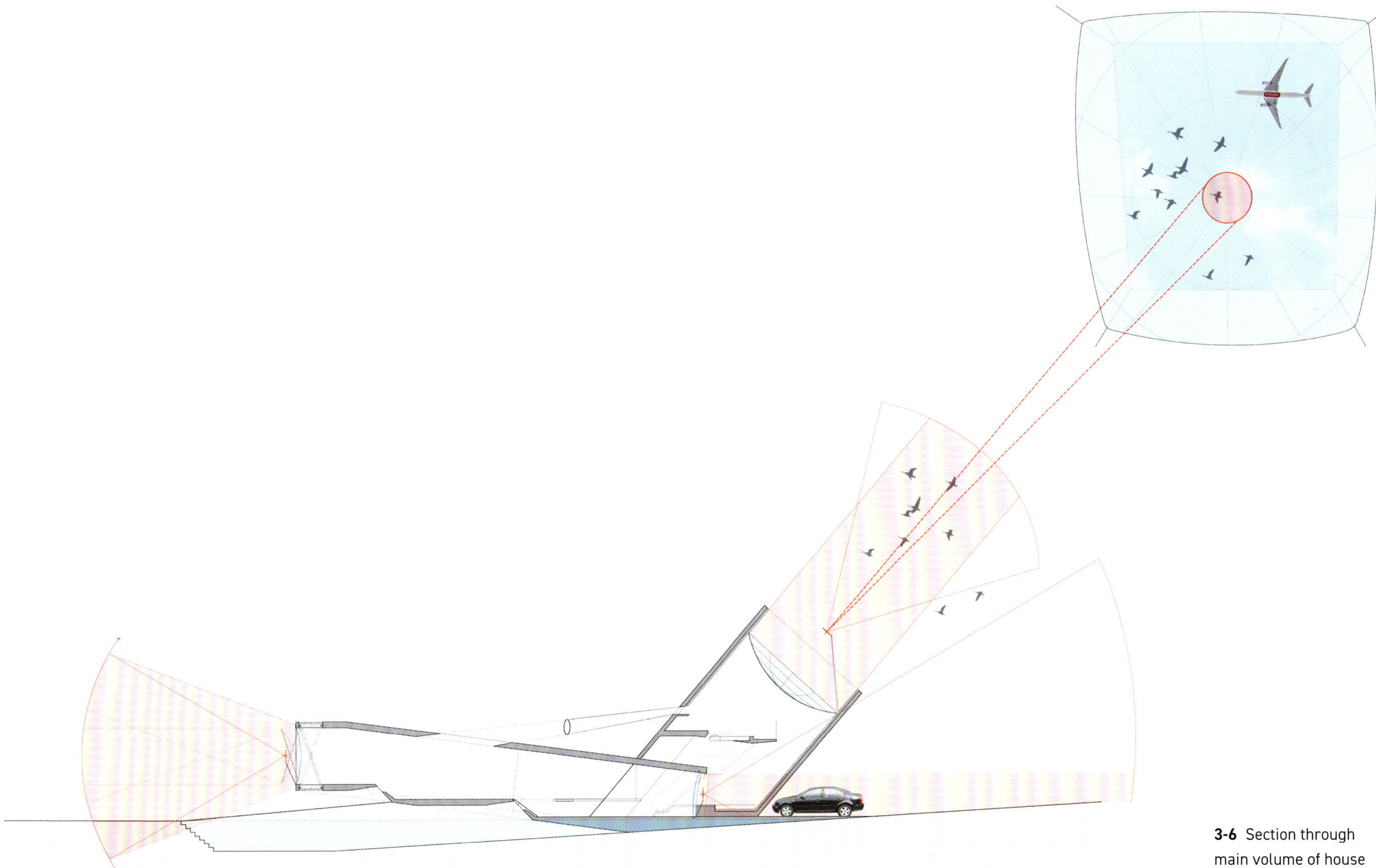

3-6 Section through main volume of house showing acoustical connection between living and bedroom areas. [Joel Sanders, Karen Van Lengen, Ben Rubin]

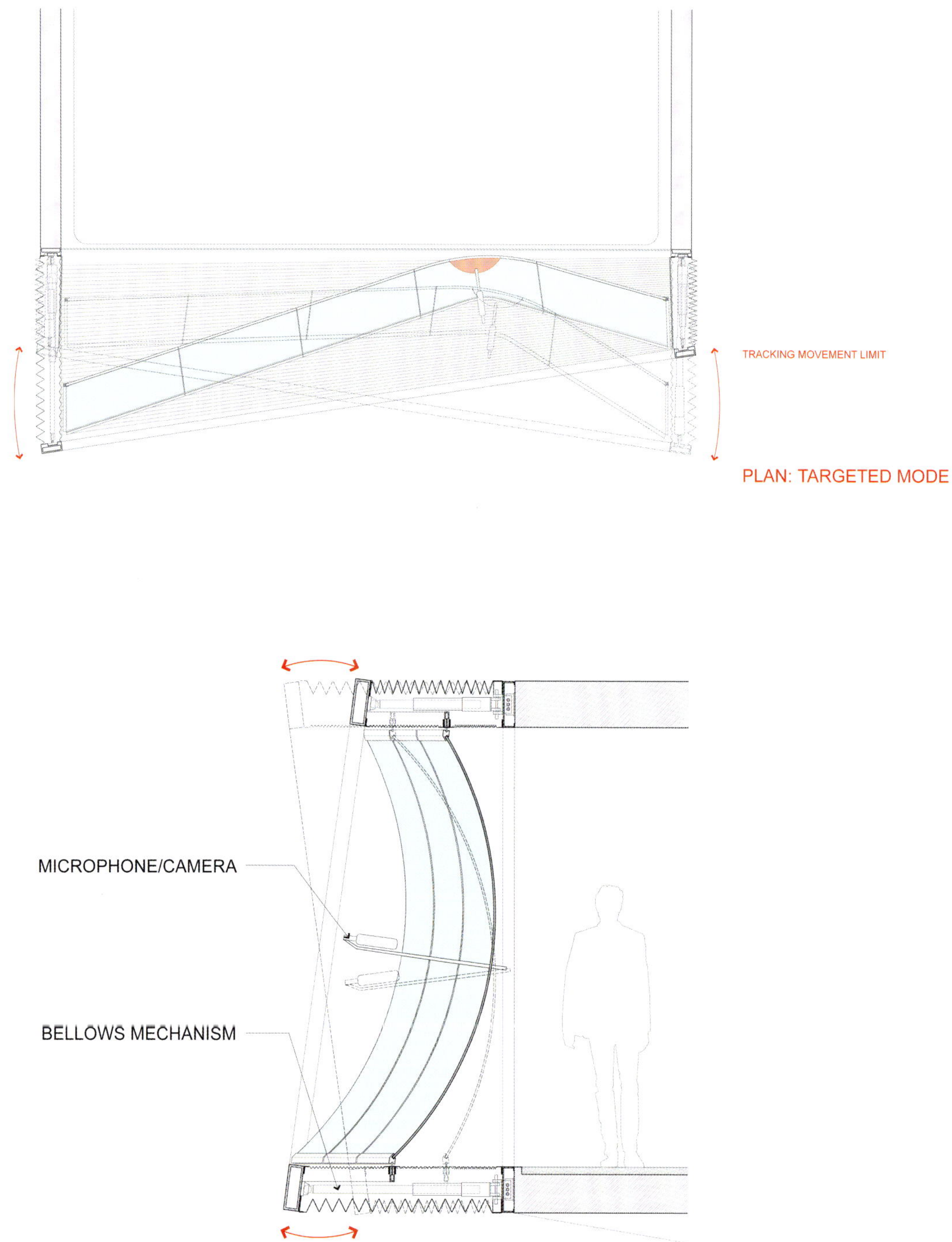

3-7 Plan and section through rotating picture window. [Joel Sanders, Karen Van Lengen, Ben Rubin]

outdoors could be mixed with Modest Mussorgsky's *Pictures at an Exhibition*; or the birds chirping in their nests might be mixed with Philip Glass's score for *Koyaanisqatsi*; or front yard and backyard sounds might be combined together with the ambient noises of the neighborhood.

This project suggests a new direction for domestic space, one that uses sound to inform its spatial design. Providing a way to transform the one-dimensional picture window into a more complex transparent apparatus that records both the visual and aural environments, it brings specificity of place to the domestic space. By creating a condition in which the occupants can orchestrate this dynamic installation, we speculate that the act of listening will not only be accentuated and coordinated with vision, but will also serve as a source of creativity and entertainment.

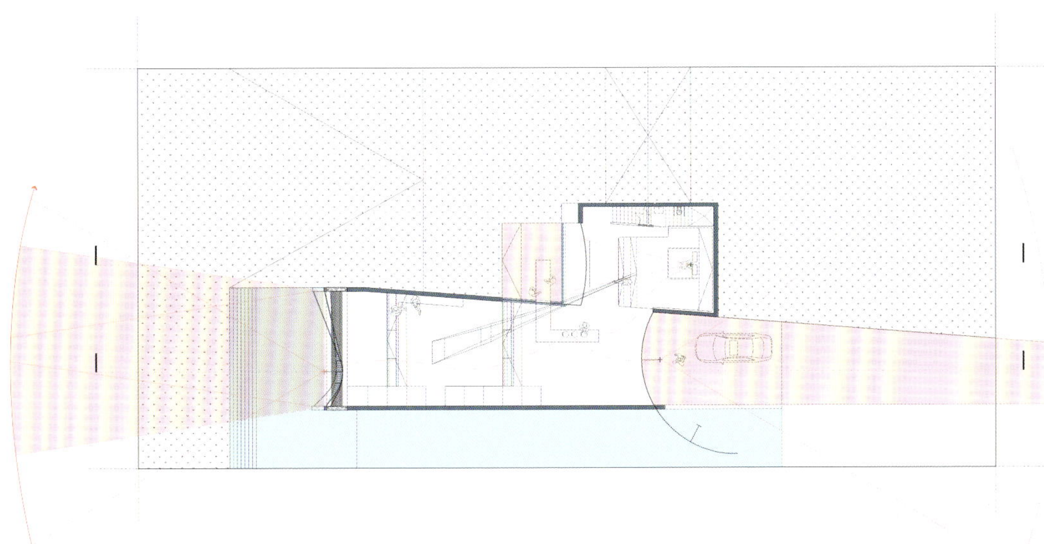

3-8 Main floor plan. [Joel Sanders, Karen Van Lengen, Ben Rubin]

Commentary on MIX House

Barry Bergdoll, Curator of Architecture and Design
Museum of Modern Art (2008)

Karen Van Lengen, Joel Sanders, and Ben Rubin have engaged in a research project that is at once timely and startlingly original. While the ideas of the relationship between sound and vision, between image and music, are deeply engaged in the history of architecture—from the ancients' interest in harmony and eurythmics to medieval and Renaissance ideals of proportions in the modern period—this relationship has largely been an instrumental one of the science of acoustics, a science progressively detached from the larger philosophical and cultural practices and ideas of the *Gesamtkunstwerk*, towards simply the optimization of sound quality for buildings functioning as performance spaces. There are but a few noted exceptions, such as Le Corbusier's late collaboration with the musician Iannis Xenakis on the Philips Pavilion in Brussels in 1958.

As practicing architects, Sanders and Van Lengen critically engage with both the tradition and its effacement through design. *MIX House* addresses not only this larger historical problem, one that is stated and deployed with a stunning originality that invites us to rethink the past, but it also raises questions about contemporary digital culture in fresh and provocative ways. As a design project, *MIX House* recalls the earlier work of both architects with domestic space as visual mediation but extends it now, not only to a different sense, but also to a whole new register of cultural interrogation.

MIX House stands alone as a design proposition but it also provides, in the tradition of many manifesto designs, the starting point for a discussion of great resonance. It is in this sense both a scholarly and a critically engaged design intervention with the potential to be remembered as a seminal statement of a new configuration of cultural issues in a still emerging digital culture with enormous implications for the social and political realms of daily life.

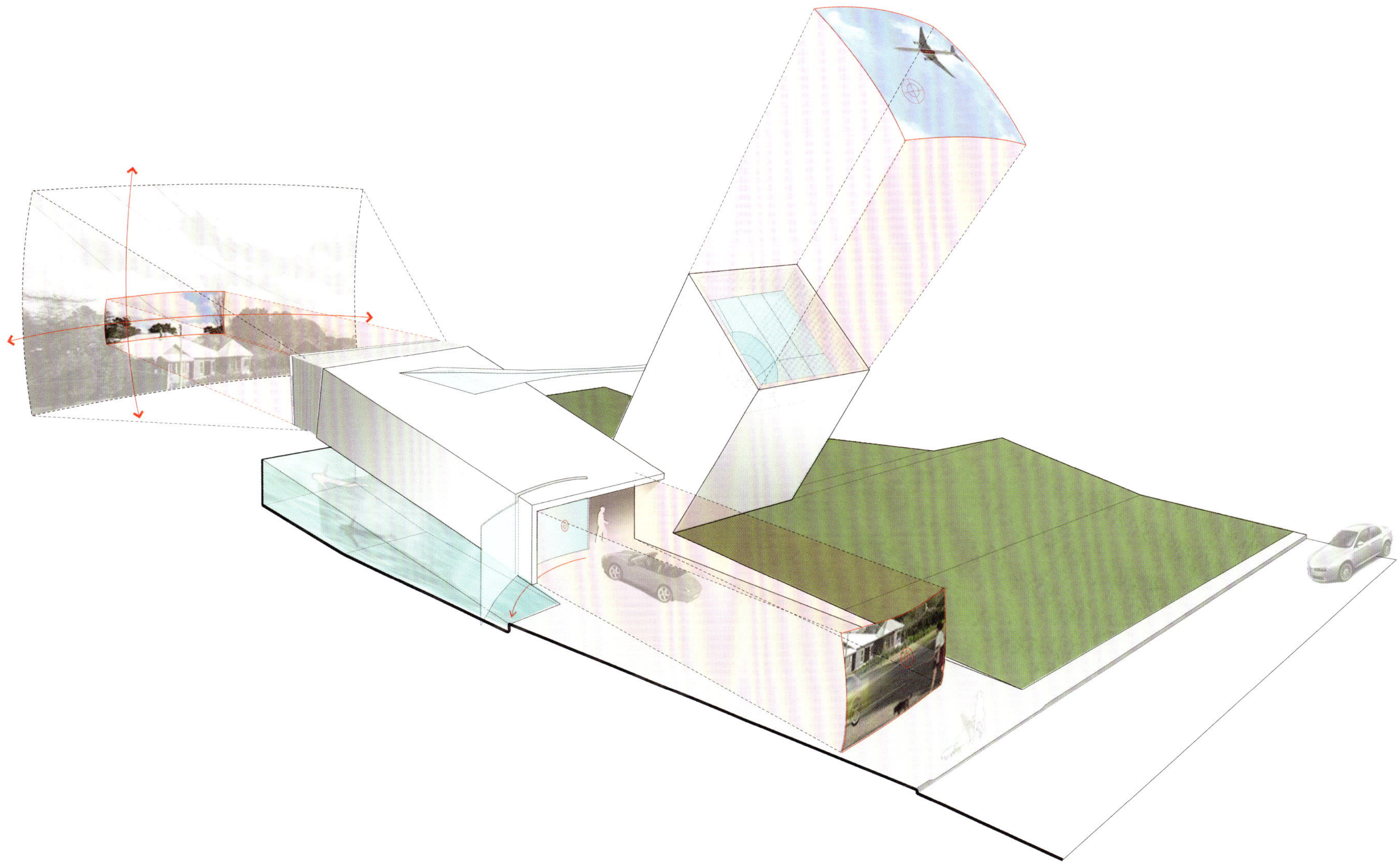

3-9 Axonometric view as seen from the street. Three sonic windows open to different landscapes, affording unique opportunities for sound transmission. [Joel Sanders, Karen Van Lengen, Ben Rubin]

Communication

The problems of deafness are deeper and more complex, if not more important, than those of blindness. Deafness is a much worse misfortune. For it means the loss of the most vital stimulus—the sound of the voice that brings language, sets thoughts astir and keeps us in the intellectual company of man.[1]

HELEN KELLER

This chapter presents a reflection on how architecture intersects with the evolving landscape of communication technologies. It highlights a critical tension between the spatial and communal nature of architecture on the one hand, and the increasingly individualistic and private ways in which people engage with the world around them, on the other. Architecture, at its core, is concerned with the organization and manipulation of space to facilitate human interaction, whether public or private. This focus, in its best form, enables engagement, fosters community, and encourages shared experiences. Contemporary culture, with its constant technological advancements, has challenged historical patterns of human interaction, isolating individuals from one another. Architecture is now often in competition with emerging technologies. As designers work within this reality, they are charged to invent new spatial ideas that can successfully address these conditions.

We begin with an essay by Olga Touloumi on the design of the Woodberry Poetry Room (1952) at Harvard University, followed by a presentation of the *Sound Lounge* (2009–19) at the University of Virginia. These designs are presented here as examples of how architecture can respond to this challenge. These projects actively promote a particular mode of engagement—listening. The Woodberry Poetry Room at Harvard encouraged visitors to experience poetry in a communal setting designed for both private listening as well as group readings. Similarly, the *Sound Lounge* at the University of Virginia provided an ephemeral space that re-envisioned how people might listen together, making shared auditory experience the focal point of the design. Both projects suggest ways to restore a sense of community using architectural design to promote the act of listening, where sound becomes the medium that connects people—both to each other and to the physical spaces they occupy.

Sound in Silence: Gramophones in the Woodberry Poetry Room

Olga Touloumi

> The place from which one speaks is outside the scriptural enterprise.[2]
> —Michel de Certeau, *The Practice of Everyday Life*

In 1949, the newly built Lamont Library opened its doors to Harvard University's undergraduates. On the fifth level of the library, a dedicated poetry space held the promise to bring the spoken word inside the silent citadel of the written. The Woodberry Poetry Room, dressed in Swedish elm and birch, included four *sui generis* pillar-like listening stations commanding the center of its interior. Inside each one of those custom-made listening stations, a record player with four sets of audio sockets and an amplifier hidden beneath, awaited listeners to take their seats, plug in their headphones, and gather around the spinning turntable. Along the windows overlooking Harvard Yard, four lounge chairs offered additional listening stations, wired to the cabinets in the center. Walls lined with books made room for the veneered consoles, perforated wooden screens, an upholstered couch, and a table, which transformed the room into a modern lounge similar to the ones featured in popular design magazines or Hollywood films of the era. The Woodberry Poetry Room, as it was named, oscillated between the informality of a private interior and the formality of a public library space, between leisure and work. Low-hanging bell lights and sconces punctuating the bookshelves and floor lamps created a contemplative environment for meaningful

4-1 Photograph of Woodberry Poetry Room, 1950. [MS Am 3147, Houghton Library, Harvard University]

encounters around books and records. Cork tiles dampened any sound of walking heels, murmuring, and other noises of library life, with stools scattered around, indicating the shifting acoustics of the room. Within this small interior that was to celebrate the poetic voice, silence reigned.

The Woodberry Poetry Room, one of the first library rooms dedicated exclusively to poetry, was designed by the Finnish architectural office of Aino and Alvar Aalto, whose architecture projects and Artek furniture had won critical acclaim.[3] While teaching at MIT, the office had undertaken the design of a new dormitory for its campus, the Baker House, and Harvard followed suit with a commission for the small interior of the Woodberry Poetry Room, designed to introduce students to new media forms of poetry beyond the book and the journal.[4] This little poetry room, I argue, holds a distinct position within narratives of modern architecture for the ways in which it demonstrates that architectural modernism was aural as well as visual. It offers evidence from a time when gramophones and turntables declared a place not only in the library, but also in architecture's imaginary constitution.

Geographies of the Modern Library

In the early 1940s, Harvard activated a pre-existing plan to expand its library system and its growing collection. Along with two new facilities, one for rare books and manuscripts, another for a book depository, Harvard dedicated a third building, the Lamont Library, to undergraduate education. For the design and construction of the new library, Harvard appointed Coolidge, Shepley, Bulfinch & Abbott, the main architectural firm building on campus at that time. The new building at the Harvard Yard would be "modern," stripped of ornament and other revivalist particularities, and presented to the world as a material mass.[5]

Unlike other research libraries, Lamont's undergraduate library made its collections more "readily accessible to the students."[6] Inside the new library, conference rooms, alcoves and carrels structured different scales of sociability around books. Scholars and administrators alike praised the library for the "absence of red tape" and "freedom of access."[7] The new building also allowed administrators to clear Widener—Harvard's main library—of the old Poetry Room, which had opened its doors to the public on May 29, 1931, and which was commissioned by oil heir Harry Harkness Flagler in memory of his beloved mentor, the Columbia University professor and poet George Edward Woodberry.[8] Harvard administrators moved the Poetry Room, along with its sister Farnsworth Room, from Widener to Lamont. In the Lamont Library, the architects united the two sister rooms on the fifth level of the new building, adding a third one called the "Forum." These three rooms ensconced a hub for both socialized learning and study in respite. An internal corridor allowed one room to bleed into the next in an acoustically porous environment that kept the interiors connected yet visually and spatially separated from one another.

Recording Voices

What distinguished the Woodberry Poetry Room from other reading rooms was sound. The poet's voice holds an almost mythological place within the production and circulation of poetry. Robert Frost, who recorded for Harvard, had called the ear "the only true writer and the only true reader," claiming that "eye readers … miss the best part of what a good writer puts into his work."[9] But inside the Poetry Room, the poetic voice claimed central stage in the form of shellac gramophone records.

Frederick C. Packard, professor of public speech, who masterminded the Harvard Vocarium, the recording label associated with the Poetry Room, had initially hoped to use gramophone records for the study of accents and elocution. When, however, Packard's "library for the voice" moved to the Poetry Room at the suggestion of Harvard administrators, the very nature of the collection changed as well.[10] By the late 1940s, the Harvard

Vocarium featured an impressive list of poets, ranging from T. S. Eliot and Ezra Pound to Muriel Rukeyser and Elizabeth Bishop. Some poets criticized gramophone records for transfixing the poem in the poet's voice, limiting the variety of readings possible. Others argued the opposite, noting that recordings interrupted and varied the inner voice of the reader rather than reinforcing the presence of the author. T. S. Eliot, who recorded for the Harvard Vocarium, declared the agonistic space between the inner voice of the reader and the projected voice of the gramophone "pregnant with interpretative potential."[11]

The Poetry Room created new social grounding for the commodification of the voice of the poet. While poets were debating their value and utility for poetry, the recordings inscribed the poet's voice within a new economy of knowledge circulation, detaching it from its "social contexts" and reintroducing it as a commodity to be collected, catalogued, exhibited, and circulated.[12] In its first installation at Widener, the Poetry Room anticipated a future of "private" listening stations similar to the ones encountered in record shops, to secure the conditions for concentrated listening.[13] Packard found collective listening to be susceptible to distraction, a popular criticism of radio at the time. Instead, he promoted an individuated kind of listening that would allow collectives to listen to poetry, echoing the architecture of music parlors, where individual stations and headphones created the conditions for attentive listening. This is the audile technique that Jonathan Sterne argues has created a private acoustic sphere enacted and shared in public.[14] That audile technique would become the core of the new poetry room, renovated and reinstalled inside the Lamont Library.

Designing for the Ear

The initial plans for the Poetry Room by Coolidge, Shepley, Bulfinch & Abbott delivered a room for browsing periodicals and books. Echoing plans for music libraries at the time, they had proposed a series of open browsing stacks, with a small number of private listening stations tucked away, subordinating listening to the operational structure of reading.

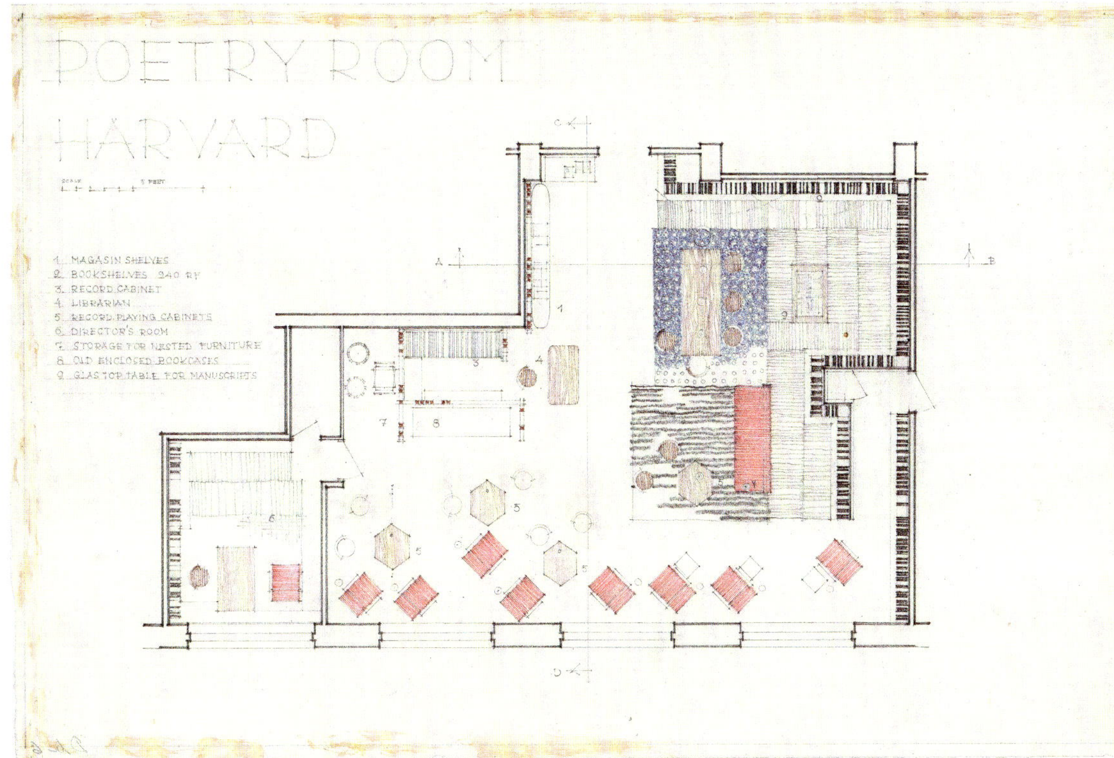

4-2 Sketch plan of Woodberry Poetry Room by Aino Aalto, c. 1947. [Alvar Aalto Museum/ Artek Collection]

Aino and Alvar Aalto, however, had a different idea. Rather than reifying listening as a private practice, they sought to redesign a space where students would encounter the spoken word. Instead of hiding them away, Aino and Alvar Aalto designed the place of sound reproduction technologies in the new library. The two architects did not mythologize machine aesthetics and the presence of technology. Alvar Aalto had been openly opposed to the functional imperative of early modernism since the early 1940s, when he published "The Humanizing of Architecture," in MIT's *Technology Review*.[15] He argued against a widespread glorification of architectures of industrialization—factories, silos, and other typologies supporting mass production—while criticizing architectural functionalism for failing to incorporate the body beyond its mechanics. "Architecture has been rationalized mainly from the technical point of view," he noted.[16] To "humanize" architecture, Aalto believed that designers needed to reinterpret the modern user as a body with a psychology; and acoustics, already a "primarily physiological and psychological" field, could show architects the way.[17]

Inside the Poetry Room, Aino and Alvar Aalto recentered the plan around the private listening stations. Aino Aalto, who had just completed the private library of Villa Mairea, Noormarkku, Finland, delivered a textured interior that modernized the bourgeois domesticity of the old Poetry Room, leaving the ornamental work to material, color, and textures. Instead of pushing them against the wall, or hiding them within long birch consoles, they placed the listening stations at the very center of the room, celebrating them as free-standing monuments to sound reproduction. At first, the two architects deployed the hexagonal form for the listening stations throughout the plan, echoing early 20th-century attitudes toward gramophones and radio sets as the technological hearths for domestic life. In this sense, the hexagonal listening stations not only monumentalized the presence of sound reproduction but also proposed turntables to act as the metaphorical pillars of the poetry community at Harvard.

When Alvar Aalto left for Finland in 1948 to tend to Aino's declining health, the Poetry Room was taken over by his assistant Olav Hammarstrom. Most of their correspondence during that period focused almost exclusively on the "gramophone cabinets," their shape, exact dimensions, and placement in the room.[18] Gradually, the shape of the listening stations changed from hexagonal to chamfered pillars, further reducing the number of simultaneous listeners a station could accommodate.

Gramophones played a special role within the work of the office, and especially so in Alvar Aalto's writing. Alvar was fascinated with sound as another material of modern architecture. He believed the gramophone to be yet another media prosthesis for the body, acting as the mechanical mouth of modernity. In fact, he purchased one right around the time he was reinventing himself as a modern architect.[19] For him, gramophones constituted new types of "objects" endowed with "cultural values."[20]

Aalto was not the only one to declare sound reproduction technologies the vital components of architectural modernism. At the same time, the Scandinavian modernist group *Acceptera* stated that gramophones promulgated mass entertainment and that the radiophone constituted a tool for cultural imperialism with a civilizing mission to bring "culture" to the rural countryside.[21] These new-at-the-time media were changing the private and public landscapes, and both Aino and Alvar Aalto were taking notice.

The Aaltos had already engaged with acoustics and the question of sound in their own work. In 1922, maybe in response to the acoustic mirrors that the British Army used along the coasts of the UK, Alvar Aalto designed parabolic acoustic shells for open door concerts at the Tampere Fair in Tampere, Finland. In 1929, pavilions with neon signs reading "radio" featured centrally at the master plan of the 700th Anniversary Exhibition and Trade Fair in Turku, Finland.[22] The Vallila Church in Helsinki gave the Aaltos the first opportunity to combine radial acoustic models with corresponding architectural forms. In the Taulumäki Church Competition Entry in 1930, the two architects designed radial plans for acoustics with multiple vaults to diffuse sound.[23]

More importantly, the Aaltos had already dealt with the problem of library acoustics at the Viipuri Library in Vyborg, Russia, one of the office's most celebrated projects. For them, acoustics and attention to the "human ear" were critical to the design of libraries.[24] To illustrate the importance of sound, the Aaltos even furnished the lecture hall with a monumental metaphorical wave of almost 30,000 Karelian pine strips, giving both architectural and acoustic form to the act of transmission. The wave in the library stood for an acoustics of distribution, a democratic auditorium that diffused the spoken word regardless of the position of the speaker. "My acoustic construction is aimed at making every point in the auditorium equal as a transmitter and a receiver of words spoken," Alvar Aalto argued, assigning a radio reality for his auditorium, where the undulating ceiling not only equalizes sound transmission but also effectively enables it.[25]

Although acoustically the metaphor did not quite work—diffusion works against speech intelligibility—architectural historians and critics of the period were more than willing to buy into the library's pseudoscientific trope. Comparing the curvy form to Miró's "serpentine lines," Sigfried Giedion claimed the roof to be the harmonious merging of "scientific reasoning and artistic imagination" minus the "rigidity" of functionalist architectures of the past.[26] Kenneth Frampton praised its acoustically considerate design,[27] and Ada Louise Huxtable noted that, once separated from its aural *raison d'être*, this undulating form became a signature profile of the Aaltos' modernity.[28]

For the Poetry Room, the Aaltos moved away from formal explorations of acoustics and focused on designating geographies of listening within silence. In fact, all listening at the Poetry Room was to happen in silence. Since books required for Aalto "a strange kind of concentration," architecture's "duty" was to "eliminate all disturbing elements," amongst them noise.[29] Apart from public readings, any other engagement with poetry was to happen privately in this acoustically insulated environment. For Alvar Aalto, acoustic insulation was central to the density of modern life. He argued that insulation against noise, be it of the city or of neighboring flats, allowed for more intimate groupings of people.[30] For him, the modern city, with its densely packed yet separate socialities, was predicated upon the architectural capacity to insulate. Silence was also healing. At the Tuberculosis Sanatorium, the two architects soundproofed the walls and designed new washing sinks so that "the water runs soundlessly from the taps."[31] Silence made an acoustic environment modern, simply by signifying the presence of resources and techniques to keep the all-pervasive din of modern life outside one's place.

When the new Poetry Room opened its doors in February 1949, this exact dialectic would play out, with the chrome and plastic of recording technologies rubbing against the wood, cork, and plywood of the Aaltos' design. The chamfered listening stations were evidence of the Aaltos' imperative to "humanize" technology, through the act of design.[32] A decade later, their listening station made one more appearance inside Vuoksenniska Church in Imatra, Finland, this time to mark the site of theocratic transmissions, the pulpit.

Conclusions

The Woodberry Poetry Room tells the story of how multiple actors shaped one of the key interior spaces of modern American poetry during the interwar period. From Harkness Flagler's parlour to Aino and Alvar Aalto's mid-century listening space, the Poetry Room offered opportunities to enact listening cultures in architectural ways. As the turntables gave way to magnetic tape players, and the phonograph records to reels, the room remained continuously a social space for poetry, tucked away, yet much alive with the spoken word.

The Poetry Room shows the multiple ways in which architecture participated in shaping listening practices and cultures, beyond the concert hall. Scholars often argue that the post-electrification acoustic modernity of transmitters and amplifiers, Celotex, and acoustic tiles brought an end to acoustics as an architectural problem, substituting architecture with material science.[33] However, paying closer attention to how architects such as Aino and Alvar Aalto engaged with the question of acoustics and sound reveals the ways in which architects expanded what it means to design with sound beyond acoustics. Reverberation and soundproofing constituted a major architectural concern, but not the only one. Architects took on the design of listening cultures at large, forcefully claiming their role in reordering the acoustic register, starting with interiors.

Within the Poetry Room in particular, the investigation for a 'sound' modernity took the form of an inquiry into how one is to encounter the spoken word. Aino and Alvar Aalto brought into this project their own critique of technology and theorization of sound. Instead of appropriating the spatial arrangement that the entertainment industry had been utilizing in record shops, they decided to reinvent the wheel. At the heart of their vision were the chamfered listening stations that emerged as the centers of social life around poetry. In fact, they monumentalized the presence of the spoken word in the room.

Sound Lounge

Karen Van Lengen, Joel Sanders, Jim Welty

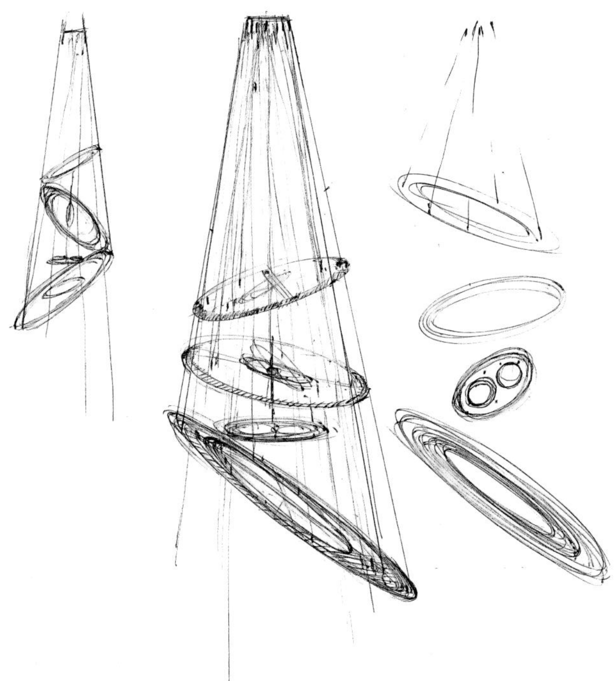

4-3 Sketch of listening cone design for *Sound Lounge*, 2008. [Jim Welty].

Karen Van Lengen, Joel Sanders, and Jim Welty designed and built *Sound Lounge* (2009–19) at the School of Architecture at the University of Virginia, pushing the Woodberry Room precedent to another level of engagement in an institutional public realm. Taking as a point of departure the technologies that separate us in our daily lives, this project integrated architecture and media through the installation of three "sound showers" that encouraged students and professors to share their audiovisual experiences. Its architectural vocabulary was ephemeral in nature, as it created rooms defined not by walls but by sound.

This renovated student lounge located at the central opening to the School of Architecture, integrated state-of-the-art audiovisual technologies to transform the formerly dreary double-height lobby space into an interactive public realm.

The design team used three audio-focused speakers to direct narrow beams of sound to the spaces below. This specific speaker technology uses a transformative panel to convert the sound into very short sound waves that can then be directed only to one distinct area, rather than longer waves that will deflect and travel throughout a space. These speakers are designed to be heard by only a few people in one small specific zone. In *Sound Lounge*, they were hung from the coffered concrete ceilings; two were enclosed in double-height fabric cones that created the listening spaces for the sounds, while the third speaker was embedded in its own housing, suspended in front of an oval panel in the carpet cut-out on the nearby wall. Together, the cones and ovals defined these "sound showers" as listening areas within the large open space. The speakers focused the sound output directly down to the blue oval carpet spaces delineated on the floor. Standing or sitting under the cones offered the users the opportunity to select their own audio choices that could be shared with others who might join the space. Sitting on comfortable mobile furniture, students could tap into one of four channels that connected them with a range of programming, from the local to the global. The largest cone married sound and image, in

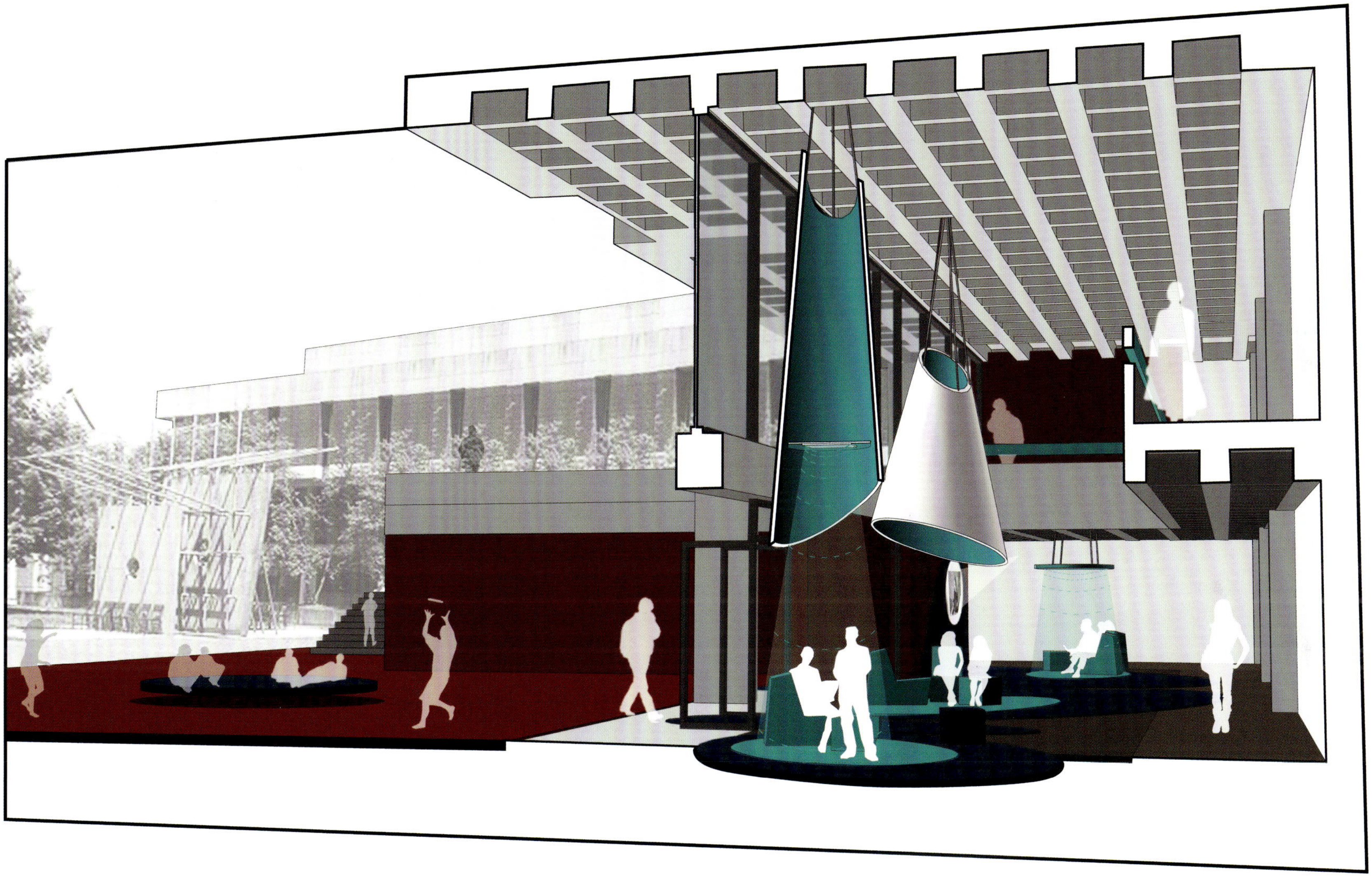

4-4 Sectional perspective showing hanging sound cones, 2009. [Joel Sanders, Karen Van Lengen, Jim Welty]

which a projected image was framed by the elliptical screen carved out of the folded carpet wall surface. This application was particularly useful when tapping into global lectures where visual materials might be presented along with the audio. Students could also use this particular space to present their own work that had significant audio components activating visual proposals. In addition, students used the sound cones during parties to play different musical scores associated with each of the three speakers while also projecting images onto the hanging cones.

4-5 Photograph of *Sound Lounge*, looking east. [Scott Smith]

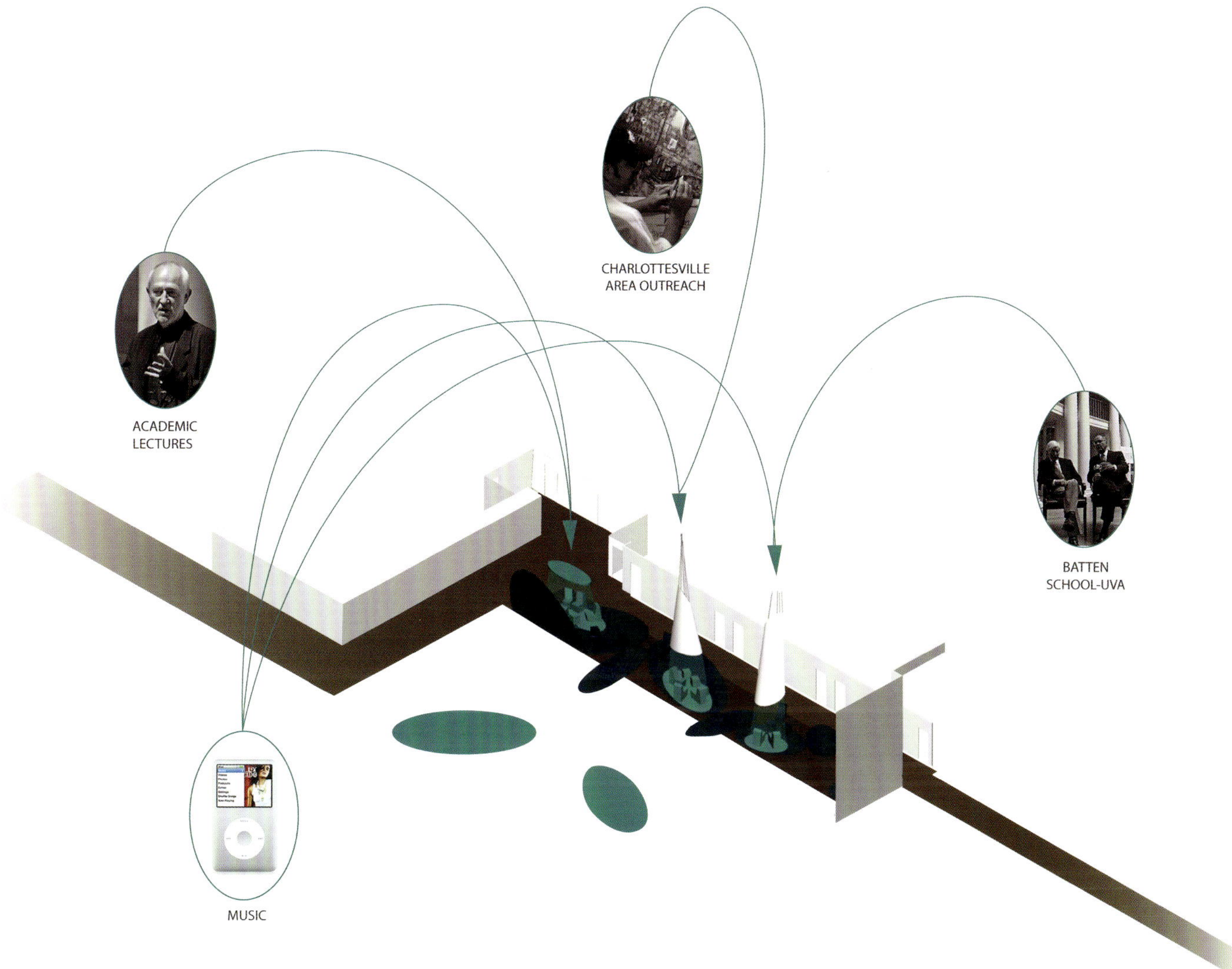

4-6 Drawing showing multiple options for sound inputs of local, regional, and global communications. [Joel Sanders, Karen Van Lengen, Jim Welty]

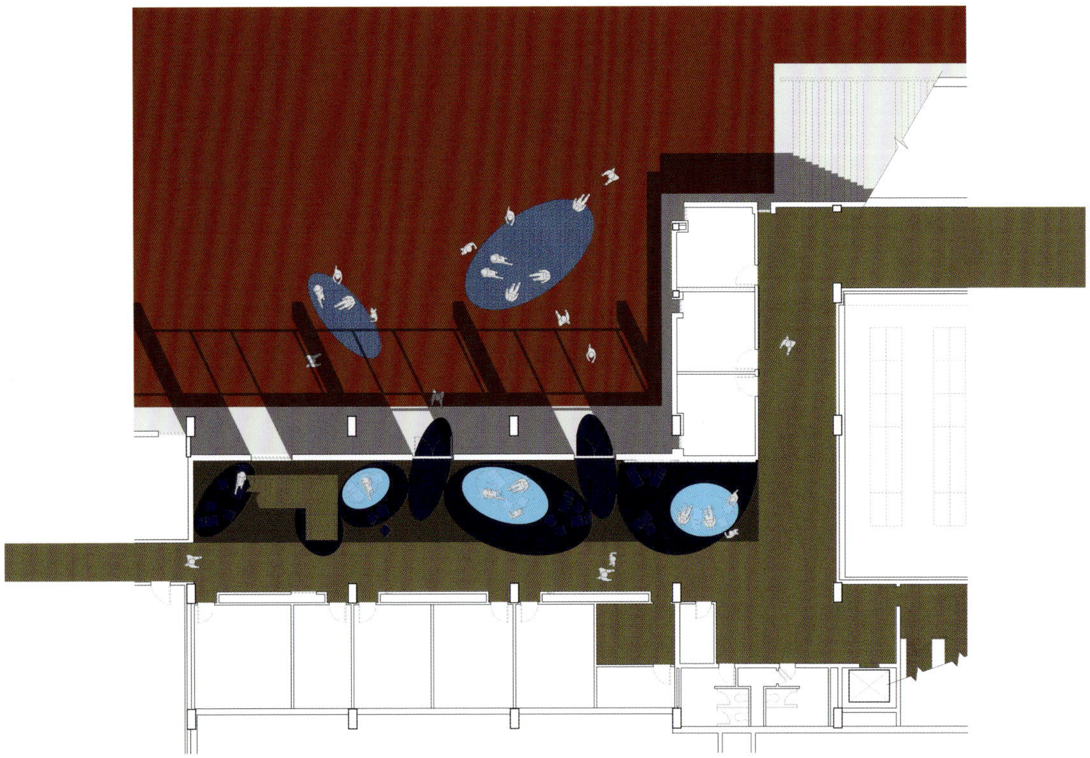

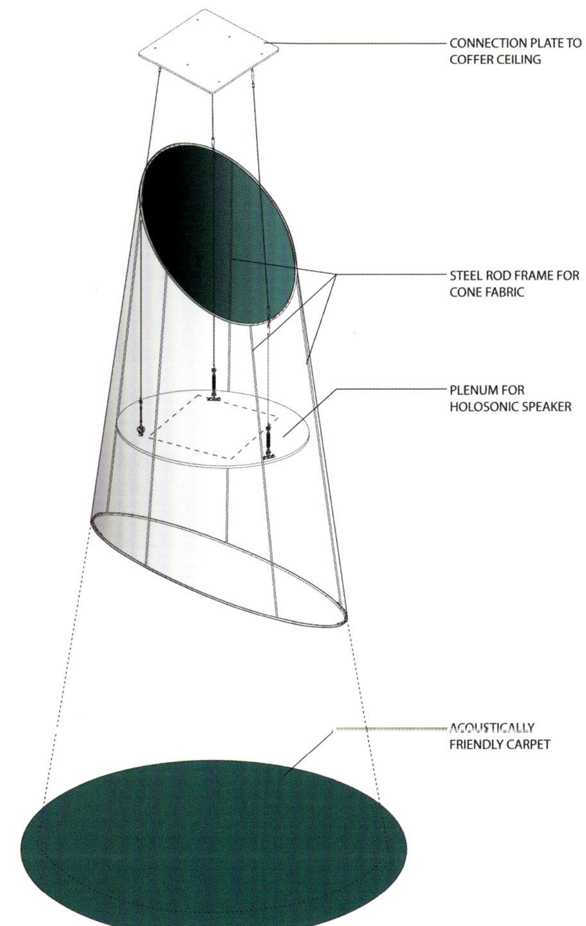

4-7 Floor plan. [Joel Sanders, Karen Van Lengen, Jim Welty]

4-8 Sound cone drawing. [Joel Sanders, Karen Van Lengen, Jim Welty]

4-9 Group review of student architecture projects that incorporated sound, 2013. [Karen Van Lengen]

4-10 Faculty and student party offering multiple sound environments using the cones of *Sound Lounge*, 2013. [Ghazal Abbasy]

Discovery

Wherever we are, what we hear is mostly noise. When we ignore it, it disturbs us. When we listen to it, we find it fascinating.[1]

JOHN CAGE

Soundscape Architecture

5-1 Drawing of Reichstag dome sounds, Berlin, Germany, *Soundscape Architecture*, 2010. [Karen Van Lengen]

5-2 Photograph of children interacting with the installation of *Soundscape New York*, Museum of the City of New York, 2015. [Richard Clarke]

Soundscape Architecture began in 2012 when Van Lengen was awarded a 2-year fellowship at the Institute for Advanced Technology in the Humanities (IATH) at the University of Virginia. Working with Director Worthy Martin and multimedia designer Lauren Massari, they developed a "sound library" of Van Lengen's sound recordings of architectural spaces amassed over many years. Out of these recordings, Van Lengen selected 20 compositions to be included in an interactive website, which showcased these different "sound spaces" along with an architectural analysis of each one. For example, the Reichstag dome in Berlin, Germany, designed by Foster + Partners, is accessed by a circular concrete ramp that winds upward along the inside glass facade of this completely transparent dome structure, overlooking the city. The shuffling of shoes and boots along this ramp provides an even and slow rhythm of movement to and from this upper realm, from where one can see all of Berlin and the sky. Here, the voices of visitors swirl in and out of auditory range, leaving notable fragments of conversations and reactions to this unusual architectural addition. In Dublin's National Library, the domed ceiling of the main reading room culminates in a linear skylight that not only illuminates the space but also provides a haven for seagulls, whose mewing calls remind visitors of the library's proximity to the sea. These, and the rest of the 20 architectural samples, were displayed on the website along an "interactive street" where visitors, by clicking on each individual wire-frame façade, could listen to a 60-second composition and review the analytical diagrams that helped to explain their sonic qualities.

The *Soundscape New York* project, which focused on five iconic buildings in New York City, grew out of this larger endeavor and was developed for an exhibition at the Museum of the City of New York. Tony Hiss, well-known author of *The Experience of Place*, spoke at the exhibition's opening, sharing his reflections on the work and the importance of sound.

Opening remarks, *Soundscape New York*

Tony Hiss, Museum of the City of New York
March 10, 2015

Thank you, Susan, Donald, Karen, and Jim. It's a thrill to be here at the opening of this wonderful show. I'm going to ask everyone to close their eyes for a moment and ask the question, "Where am I, and who else is here?"

The ears assumed their importance some millions of years ago in a post-binocular but pre-bipedal world. We had long before, while still up in the trees, sacrificed panoramic, wrap-around vision by bringing our eyes closer together and gaining depth perception. But in the tall grasses of East Africa before we stood upright, we couldn't see more than a few feet away. But we could hear in all directions for half a mile or more, just on an ordinary, everyday basis, and we could reach beyond the ordinary when necessary. Thunder is audible 10 miles away, and even today, as kids, we learn how to count the seconds between a lightning flash and a thunder roll in order to estimate how far away a storm is and how fast it is moving. Much later, after we'd settled down in villages, we still used sound to organize towns and neighborhoods. Sound cements community—a parish was essentially everyone within earshot of the bells in a steeple or the sound of a call to prayer or, in more recent times, clock towers. Early Sunday morning in my neighborhood, or during a snowfall, you can on the hour still hear competing Big Ben chimes—one set from a church on 10th Street and another, a few seconds later, from the Con Edison building on 14th Street.

Tonight, we're here to talk about uniting sound minds, sound bodies, and "soundful" buildings, and celebrate what Karen Van Lengen and Jim Welty are accomplishing. Karen and Jim are rescuing our minds, because paying renewed attention to any of the senses other than the visual brings us back into the whole sensorium (sorry about that word)—it just means everything we're equipped to take in: sights, sounds, smells, tastes, touches. Re-tuning the mind: that's what happens, whether you call it Deep Listening, as Karen and Jim do, or Deep Travel, or simultaneous perception, as I have, this is the third waking awareness—the Cinderella awareness, the one that stays in touch with the world all around us and with the never-beforeness of now. It's the experiencer, the all-channels-open alertness, that feeling of waking up while you're already awake. It's a capacity built into everyone but for too long it's been the one we haven't quite known how to incorporate into our education, since we've focused exclusively on its stepsisters, telling students to avoid daydreaming and get better at pinpoint attention.

We live at an odd moment when sound has temporarily been drowned out by noise. Ever since President Reagan in 1981 ended funding for Environmental Protection Agency noise pollution controls enacted in 1972, sound has been the only real casualty of that first era of environmental legislation: we have cleaner air and cleaner water, but we don't have clean sound. In 1973, the N/R BMT 49th Street subway station was clad in soundproofing, as an "experiment." Despite its immediate and continuing success—go take a listen—42 years later the experiment has yet to be repeated in any of the other 467 stations in the system. Noise pollution diminishes and dulls and blurs our minds every bit as much as smog smears the horizon or the smog-of-the-night, light pollution, takes away the dome of the sky, making it impossible for 80% of Americans to see the Milky Way. Since the 1980s, everything has gotten noisier, and fixing it is not even on our to-do list anymore. "Faster, better, cheaper" is an inspiring new design mantra, but it ought to be "faster, better, cheaper, quieter." The only conversations everyone can hear on the streets are those of car horns and police and ambulance sirens and fire engines, because they're the only voices loud enough to out-shout the engines and tires. And the only feelings they can express are anger and fear and urgency and "Move Over" and "Get Out of My Way." It's a different kind of 1% throwing its weight around, a throwback to the early days of the Industrial Revolution machinery, when noise meant power and progress.

So, we crave silence, which is a step in the right direction. "Silence is the new luxury" is the slogan on T-shirts given to guests at a new luxury resort in the Turks and Caicos Islands. The recently remodeled Beverly Hills Hotel has double layers of drywall to help reduce what they call "noise transfer."

But Karen and Jim are telling us we could be thinking about promoting sound, not just reducing noise. The Quiet Car on Amtrak works because, in the absence of all the cell phone calls, you can once again hear the train itself, the rattles and squeaks that let you know that you're being taken care of—and *that* gives you access to your own train of thoughts. Of course, occasionally the conductor does lean his head in to bellow, "this is the Quiet Car!"

Part of the true magic of New York City's High Line is that you can hear the sounds of the city, even birdsong, and the wind blowing through leaves and grasses, because you're lifted about two-and-a-half stories above the surface blanket of vehicles down in the streets.

Flickering, tiny screens in front of our faces and headphones and earbuds—I see them all as a desperate cry for help, a fierce denunciation of what we've let happen to buildings and to a public realm that doesn't look good or sound right.

Too often we only think about sound in designing specialized settings: libraries, classrooms, concert halls. And even there we often miss the mark. I remember my father telling me how, around 1960 when Avery Fisher Hall was under construction—the home of the New York Philharmonic that soon, after yet another acoustical renovation, will become David Geffen Hall—he had met an acoustician who told him it was a very dull field to work in because all the problems had already been solved.

Sound is amazing—it can flow like water, as older civilizations who paid attention knew. At the ball court at Monte Albán, for instance, outside Oaxaca, the sounds of the valley far below flow straight up that hill to surround the ball court. How did they make that happen? Karen and Jim are telling us that every building can have its own distinctive sounds. Grand Central Terminal even has two sets: the echoing, murmuring Grand Concourse and the secret Whisper Zone down by the Oyster Bar.

There's no time to lose—there are so many existing buildings to retrofit, and so many new buildings on the drawing boards. Let's put these buildings on the listening boards, not just the drawing boards. Let's mainstream the teaching of sound as a required design skill in architectural schools and do that right away. Let's immediately set up a checklist with performance standards for evaluating how buildings handle sound. Let's tonight establish the annual Karen Van Lengen Prize for the best-sounding building of 2015 and bestow the first award with suitable fanfare here at the Museum this time next year. It's a big 21st-century challenge, but if we get started now, it's bound to be a resounding success.

Soundscape New York

Soundscape New York, designed for an exhibition at the Museum of the City of New York, was an immersive installation that showcased five iconic architectural interiors in New York City. The installation used recorded sounds, coordinated with visual animations of each place. Jim Welty joined our team to develop these animations. We selected five building interiors: the Rose Main Reading Room at the New York Public Library, the main concourse of Grand Central Terminal, the lobby of the International Building at Rockefeller Center, the lobby of the Seagram Building, and the atrium of the Guggenheim Museum.

Van Lengen began the project by making multiple recordings of each of these spaces, while attentively listening to their distinct qualities during the recordings. Working with sound editing software, she used the recordings to create characteristic 60-second compositions for each of the selected spaces. As with the larger *Soundscape Architecture* project, she then developed analytical drawings explaining the influential factors that shape the sounds of these spaces. These diagrams included visualizations of the volume of the space, the occupancy at the time of the recording, and the materiality of each interior.

5-3 Elevation drawing of *Soundscape New York* buildings, 2013. [Karen Van Lengen]

Welty constructed the digital spatial interiors for the animations while Van Lengen made lists and then drawings of the "sound objects." These "sound objects" are the specific sounds that come and go depending on the time of day, or size and type of occupancy of the space—for example, the sounds of distinct voices, machinery, trains, or birds. They are unpredictable and always changing but are framed by the general aural characteristics of the larger space. In the development of both the drawings and animations, Van Lengen and Welty listened to these sounds in isolation, over and over, until they became just that—sound—not affiliated with a specific place or origin. They drew and animated these "sound objects," with colored-pencil drawings done by hand as well as digital drawings, which would then populate the final animations.

People often ask, how does one draw sound? Sound recordings, when transferred to auditory software programs, do have standardized forms of visualization. These spectrograms visually represent the frequencies of a recorded piece while also providing information on the volume, timbre, and other qualities of the sound. These visualizations were important in the creation of the 60-second compositions for *Soundscape Architecture*, as they provided the means to edit new compositions from the original sound recordings. However, if one wishes to imagine these sounds outside of such standard technical visualizations, there are few precedents to draw upon. Unlike centuries of drawing traditions that have documented the history of the visual arts and architecture, drawing sound is not included in the architectural canon. There are no rules. Rather, it is a highly personal and creative act. It requires invention and a desire to visually express these sounds through emotional attachments to the timbre, pitch, and volumes they project. The naming of each sound helped to convey this spirit by giving the individual sound a sense of identity. Van Lengen and Welty used these "sound objects" as "actors" in the animations that follow.

As they began this process, they studied several 20th-century precedents of early experimental filmmakers, such as Oskar Fischinger and Viking Eggeling, who had used an array of geometric forms to accompany sound sequences in creating their first filmic animations. The word animation is derived from the Latin word *anima*, which means "life" or "spirit." The animations for this project uncovered the atmospheres of these special places, using recordings and drawings to make them. Welty, a sculptor with extensive experience in set design for modern dance performances, created the theatrical ambience for

5-4 Sound drawings of the five Manhattan buildings collaged over a sketch of Manhattan, 2024. [Karen Van Lengen and Jim Welty]

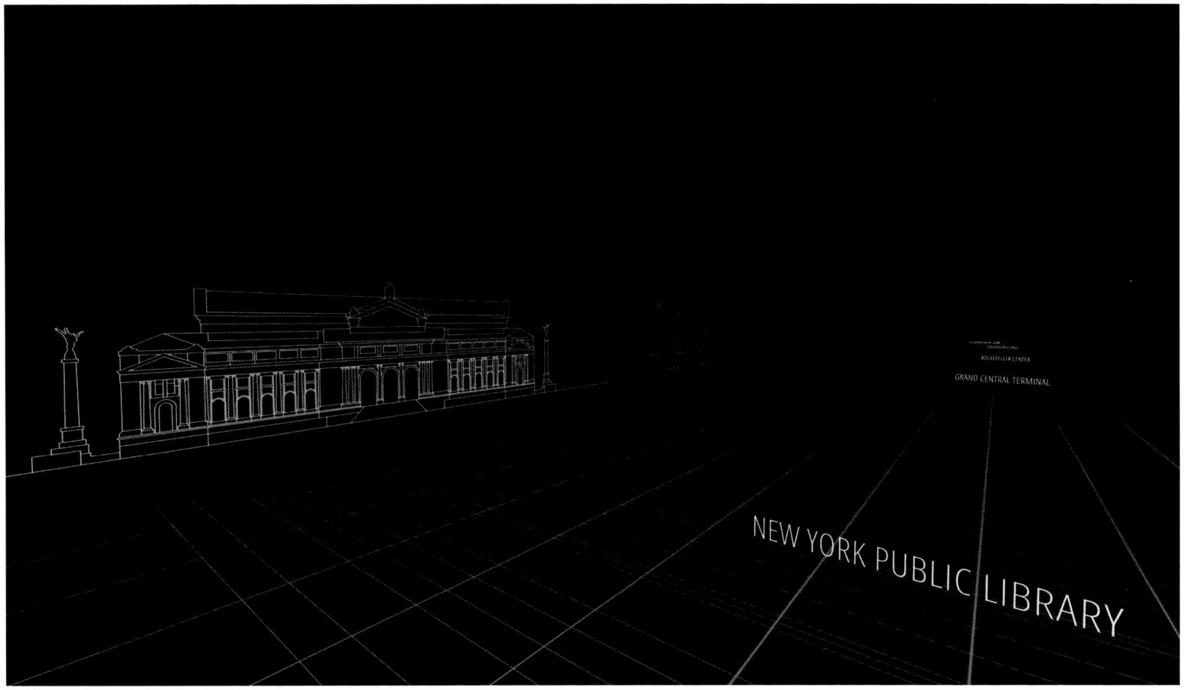

each animation. He began with photographic overlays, including photographs of the actual spaces and other graphic images that embodied their spirit. He animated these virtual enclosures, first depicting the transformations of the spaces and then conducting the sound objects to move in and out along with their accompanying voices. The resulting animations became theatrical interpretations of these interiors that we have come to know so well (see QR codes throughout to access the animations).

5-5 Digital street image of the New York Public Library, *Soundscape Architecture*, 2014. [Karen Van Lengen]

5-6 Analytical diagram of Rose Main Reading Room showing volume, circulation, and materials, 2014. [Karen Van Lengen]

5-7 Photograph of the interior of Rose Main Reading Room, New York Public Library, New York City. [Karen Van Lengen]

And finally, from these animations Welty created the last stage of this process, displayed in the following chapter, *Art*. He created digital paintings that contain new arrangements of the "sound objects" found within their preceding lives in animation form. These artworks carry the memory of those spatial encounters as they recall the spirit of these places, sonically and visually.

New York Public Library

Designed and built on the site of New York City's Croton Reservoir, the New York Public Library was established by philanthropic organizations and families in 1895, to create the main branch of an emerging library system in New York City. After the trustees ran an architectural design competition, Carrère and Hastings were selected as the architectural team. Located on Fifth Avenue, this 1911 Beaux-Arts building spans two city blocks between 40th and 42nd Streets and was rumored to be the largest marble building at the time of its opening. Its majestic size is best experienced and heard in the Rose Main Reading Room located at the top of the building. This famous reading hall has vast dimensions—a length of 297 feet, a width of 78 feet, and a height of 52 feet—creating a voluminous interior of over one million cubic feet. This enclosure includes many reflective materials, such as terra-cotta floors, marble and glass interior facades, and painted plaster ceilings. When a pair of high heels walks across this floor, one can hear the piercing footsteps distinctly. When the book retrieval system is in action, one registers this mechanical activity that brings the books from the stacks to the Reading Room. And when the heavy oak chairs pull away from the long wooden reading tables, one can hear the roar created by the friction between the two materials, a roar that harkens back to the statuary lions that frame the library's Fifth Avenue entrance.

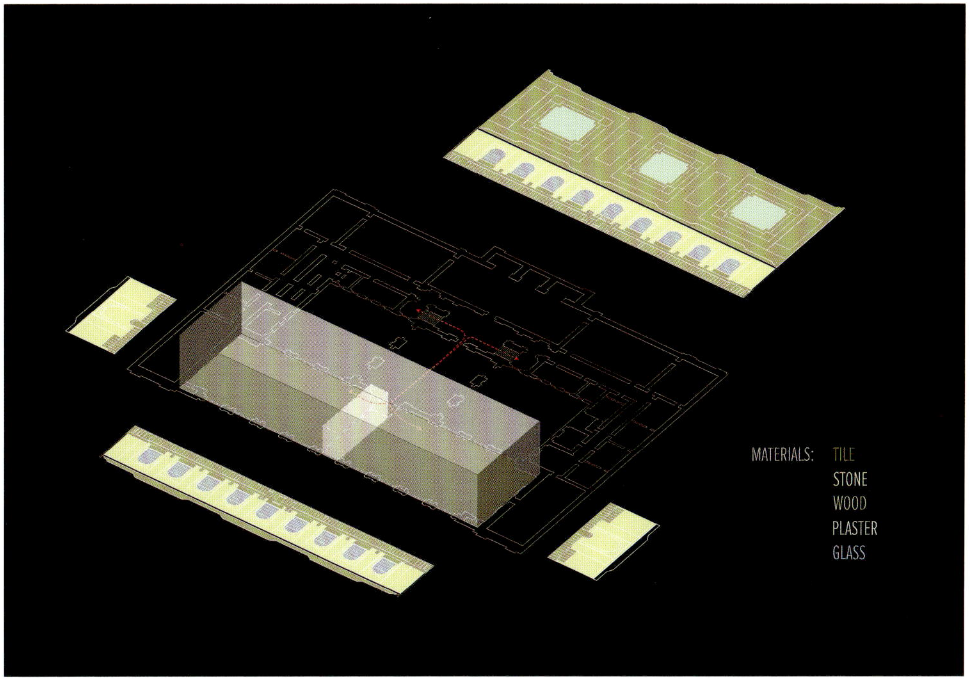

So, is the library too noisy? Van Lengen spent many hours there studying for her architectural licensing exam, sharing those characteristic sounds with the larger library community. She passed that treacherous exam and has often credited this library in its support of her mission to become an architect. Those sound memories remain familiar and part of her library experience.

Sound - Drawing Animation / Coordination
Project: New York Public Library
Date: Sept 13, 2012

Notation in seconds NAME

2.6-4.5 Slam 1

5.8-7 Sliding 1

1.2, 9.6, 11.2, 17, 101 Klink

14-16 then fade Slam 2

20.5-23 Sliding 2

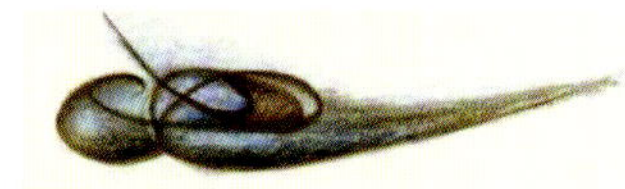

24.4-27.5 then fade Sliding 3

31.2, 33.2 Little Bang

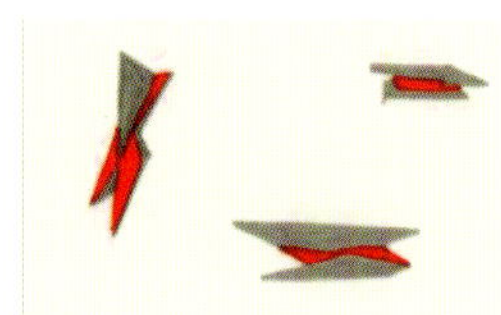

36.5-39.5 + 40, 42, 54 Deep Sliding

40, and 42 Slam Bang

58-61 Faint Slide

57 High Ring

44 to end High Heels

101 to end Voices

5-8 Process document of hand drawings associated with time markers from the sound composition for Rose Main Reading Room, 2014. [Karen Van Lengen]

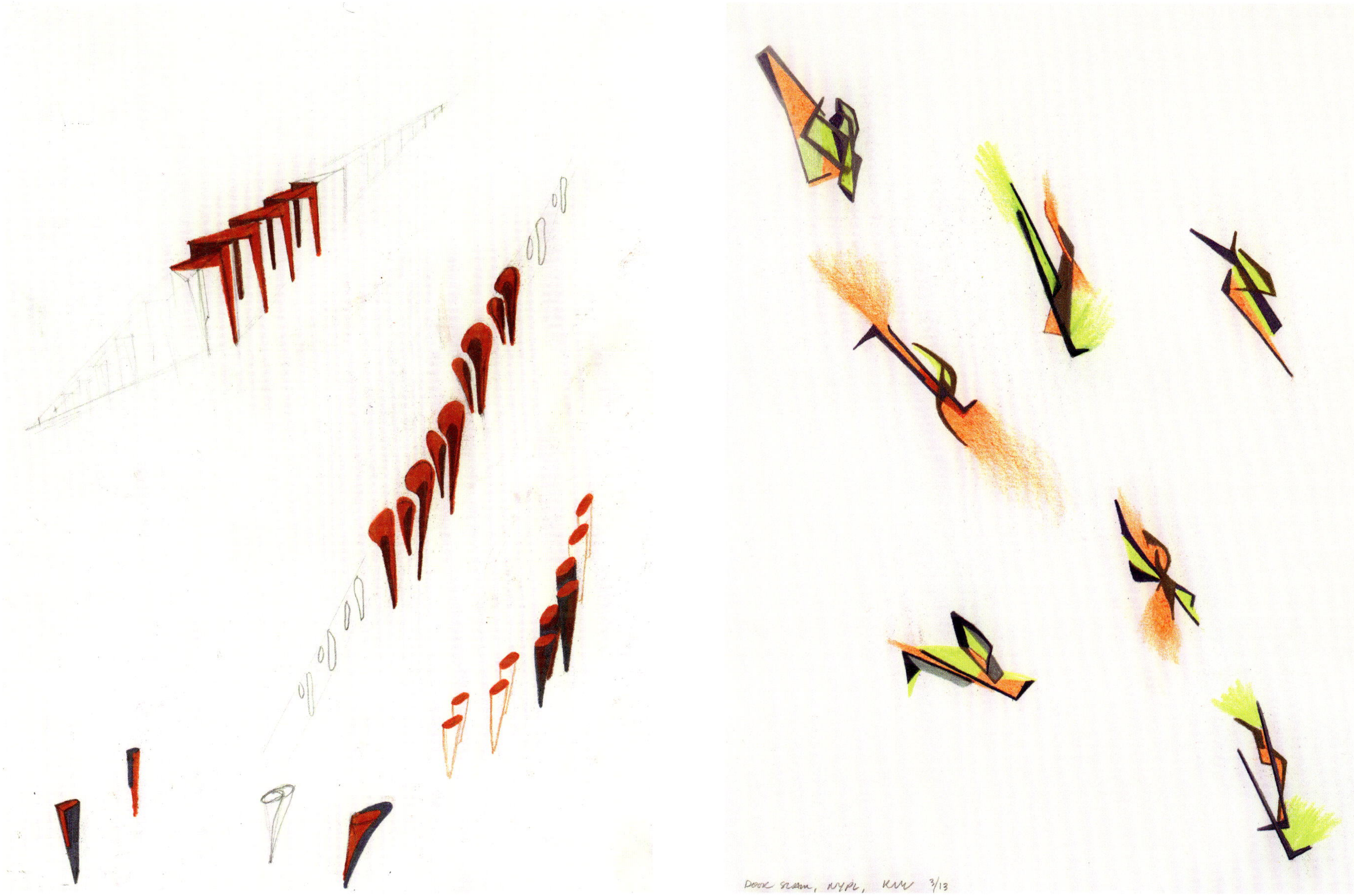

5-9 Drawings of high heels walking (left) and metal doors closing (right), Rose Main Reading Room, 2013. [Karen Van Lengen]

5-10 Digital street image of Grand Central Terminal, *Soundscape Architecture*, 2014. [Karen Van Lengen]

Grand Central Terminal

Just a couple of blocks away from the New York Public Library is Grand Central Terminal, built during the same time period and located on the site of the old Grand Central Depot built by Cornelius Vanderbilt in 1871. The new station, completed in 1913, was designed by architects Reed and Stem, with Warren & Wetmore acting as the main designers. It was an engineering miracle on multiple levels, built to accommodate the numerous transportation systems that it served, including trains, subways, and the elevated Park Avenue roadway that wraps around the hall. At its opening it was the largest enclosed interior space in the world, with dimensions of 275 feet in length, 120 feet in width, and a height to the top of the vaulted ceiling of 125 feet. The lights embedded in the barrel vault are part of the zodiac illustrations that cover the hall's ceiling. This Concourse has highly reflective surfaces, which include marble walls and floors that rise up to meet the beginning of the vaulted ceiling, finished in plastered stucco and painted surfaces. The sounds in this space, made by the hundreds of daily visitors as they are walking, running, and rolling their suitcases across the Concourse, have long reverberation times. This explains the all-encompassing, seductive, oceanic sound that we associate with this space.

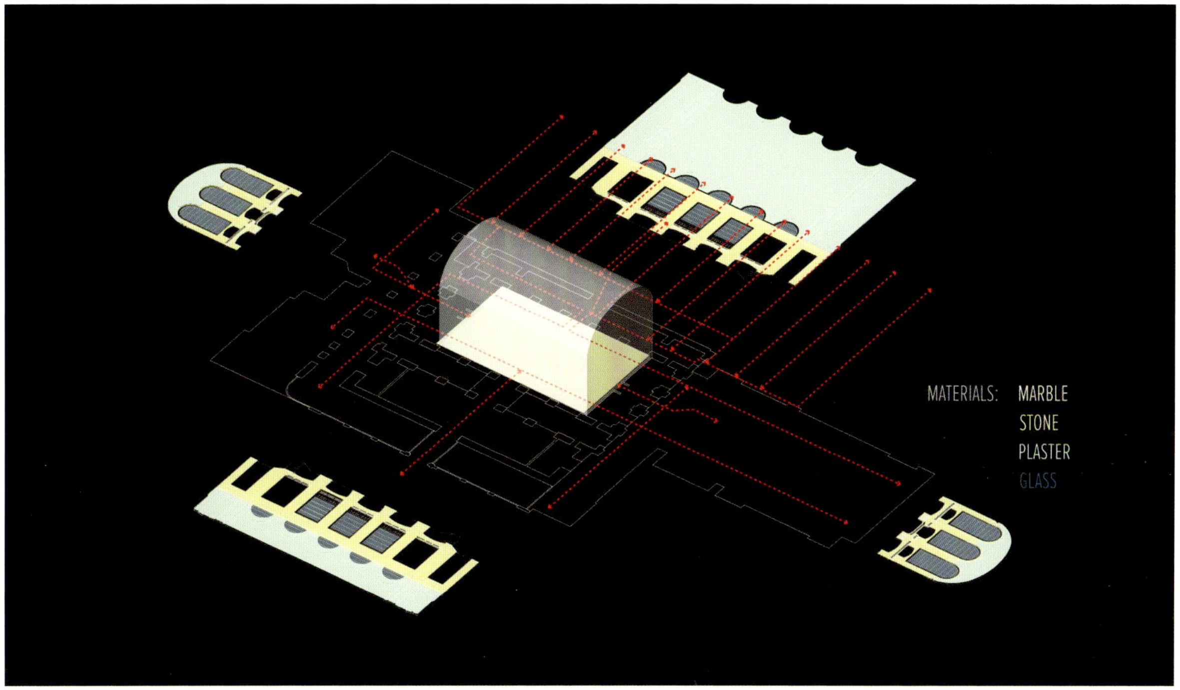

5-11 Analytical diagram of Grand Central Terminal showing volume, circulation, and materials, 2014. [Karen Van Lengen]

5-12 Photograph of Grand Central Terminal, New York City. [Karen Van Lengen]

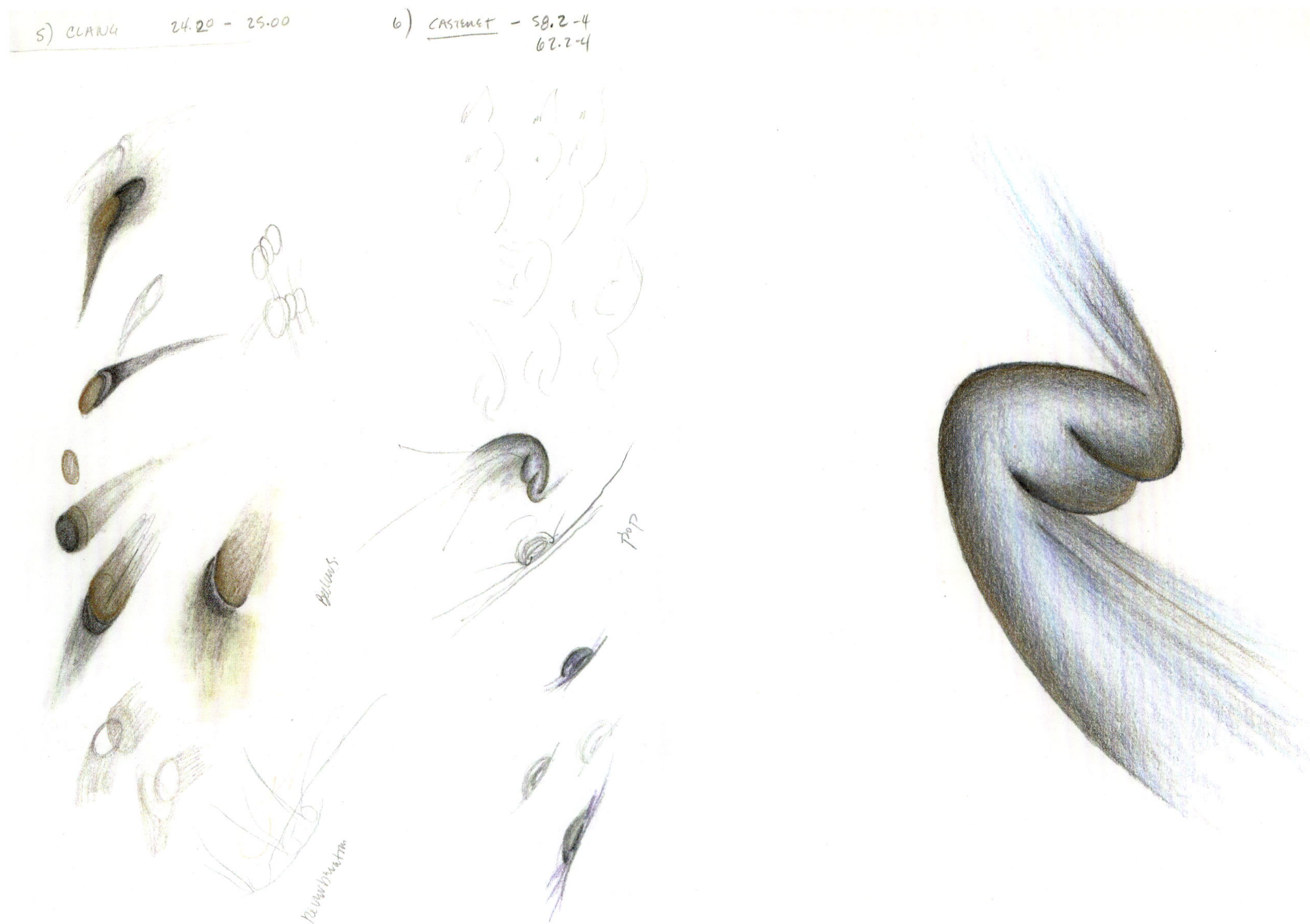

5-13 Drawings of the sounds of Grand Central Terminal, 2013. [Karen Van Lengen]

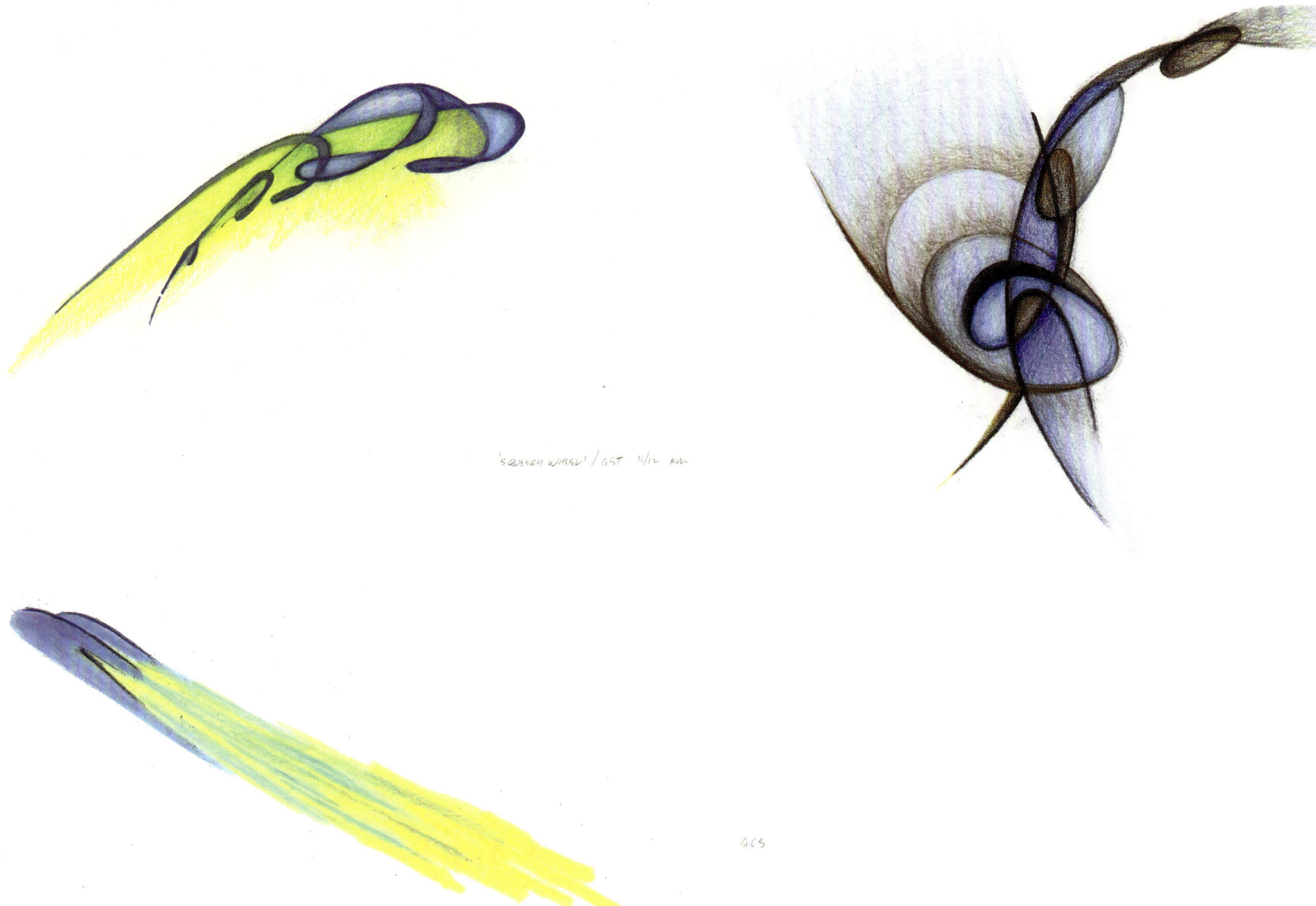
'SQUEAKY WHEEL' / GST 11/12
GCS

5-14 Digital street image of Rockefeller Center, *Soundscape Architecture*, 2014. [Karen Van Lengen]

International Building, Rockefeller Center

Rockefeller Center is a 22-acre complex of buildings and open spaces located in Midtown Manhattan, between Fifth and Sixth Avenues and extending from 48th to 51st Street. This Art Deco masterpiece was built between 1930 and 1939. A marvel of both engineering and aesthetics, it is arranged around a central sunken plaza that transforms from an ice-skating rink in the winter to an outdoor café in the summer, offering a change in program accompanied by a sonic shift as well. Its impressive connection of urban networks, including subway lines, underground shopping streets, and mid-block open spaces make it one of the most important urban complexes ever built in the United States. Designed by Raymond Hood, working with the primary architects Henry Hofmeister and H. W. Corbett, the complex includes numerous sonic spaces, from the main central lobby to the underground shopping corridor, to Radio City Music Hall, and to the outdoor sunken courtyard. However, the lobby of the International Building on Fifth Avenue has an aural quality that we found particularly interesting. This vertical lobby is sheathed in highly reflective stone surfaces that allow sound to reverberate crisply in the tall and majestic entrance facing

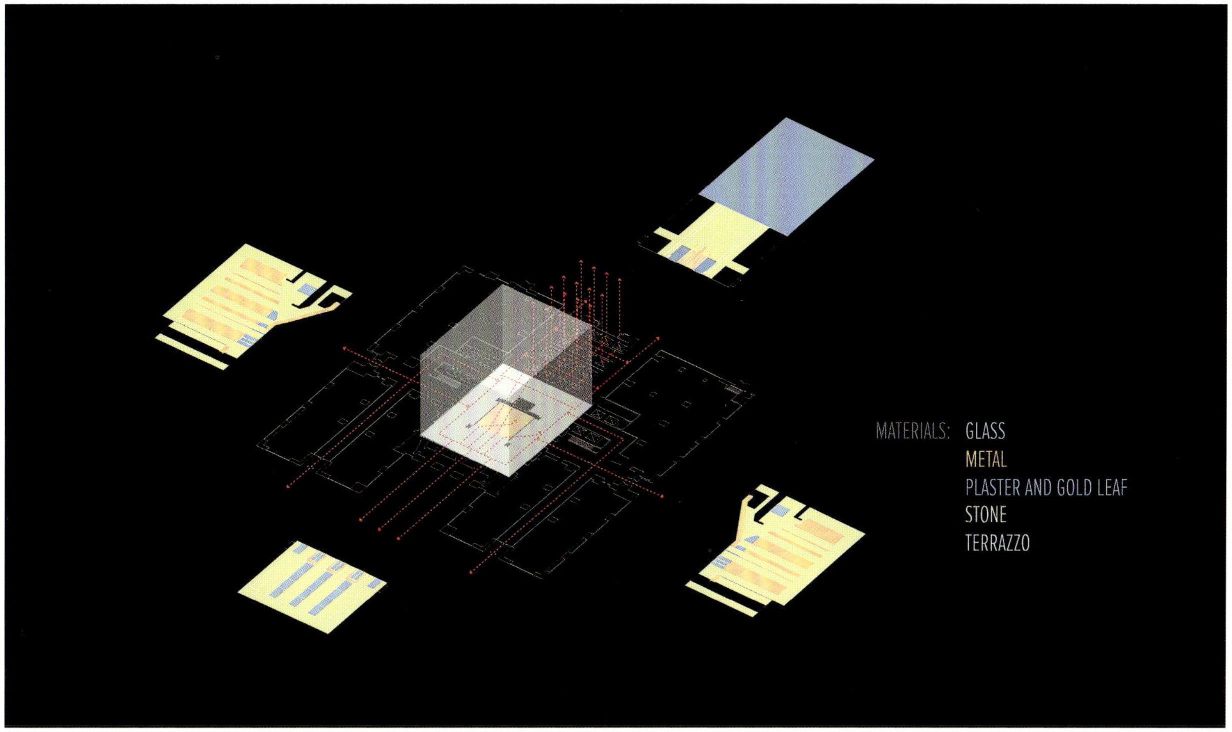

Fifth Avenue. The overall background sound includes that of the original escalators, which ferry visitors to both the upper and lower floors and serve as the main feature of this vertical space at the northeast corner of Rockefeller Center. Although landmark status was bestowed on these escalators for their unique Art Deco visual attributes, they also convey a special meaning acoustically. The audible churning of the escalators' mechanical gears creates a sound that reminds us of early industrial equipment, and perhaps specifically of the Rockefeller oil machinery and business that ultimately built this complex. In this way, the ambient sound of this space offers us a form of history, if we are listening.

5-15 Analytical diagram of the International Building Lobby, Rockefeller Center, showing volume, circulation, and materials, 2014. [Karen Van Lengen]

5-16 Photograph of the International Building Lobby at Rockefeller Center, New York City. [Karen Van Lengen]

5-17 Drawings of escalator gears churning (above) and "Oh, I'm running late" (right), International Building Lobby, Rockefeller Center, 2013. [Karen Van Lengen]

ROCK CENTER, I Know I'm running late, / KW 5/13

Seagram Building

The Seagram Building is an icon of the International Style in architecture, completed in 1958. Designed by Mies van der Rohe for the Bronfman family, owners of the Seagram beverage empire, it stands directly opposite from the New York Racquet and Tennis Club, which was designed by the well-known neoclassical firm of McKim, Mead & White, and completed in 1918. The Seagram Building is as open and transparent as the Racquet and Tennis Club is closed. This 40-story steel frame building, with a tinted glass curtain wall facade, was designed to express a reductive architectural vocabulary with a refinement in details and proportions, distinguishing it as a prime example of the International Style. The building is set back 100 feet from the street by a set of long horizontal stairs that join a granite entry podium. The stairs are bordered by two large water features whose fountains have the effect of creating a sonic gateway to the building plaza and entryway; they mask the traffic sounds of Park Avenue and begin the processional entry into the transparent lobby space. The lobby is enclosed on three sides by full-height glass walls that meet the ceramic tile ceiling. The granite floors extend directly out to the plaza outside, expanding the visual impression of this interior. The banks of elevators on the lobby's east side are sheathed in Roman travertine. The space is distinguished sonically by its proportions and highly reflective materials. The calling out of a name or the power-walking heels that penetrate the space are so distinctive and crisp that they may sometimes remind us of the racquet sports going on in the club across the street. These sounds are sharp, clear, and immediately reflective.

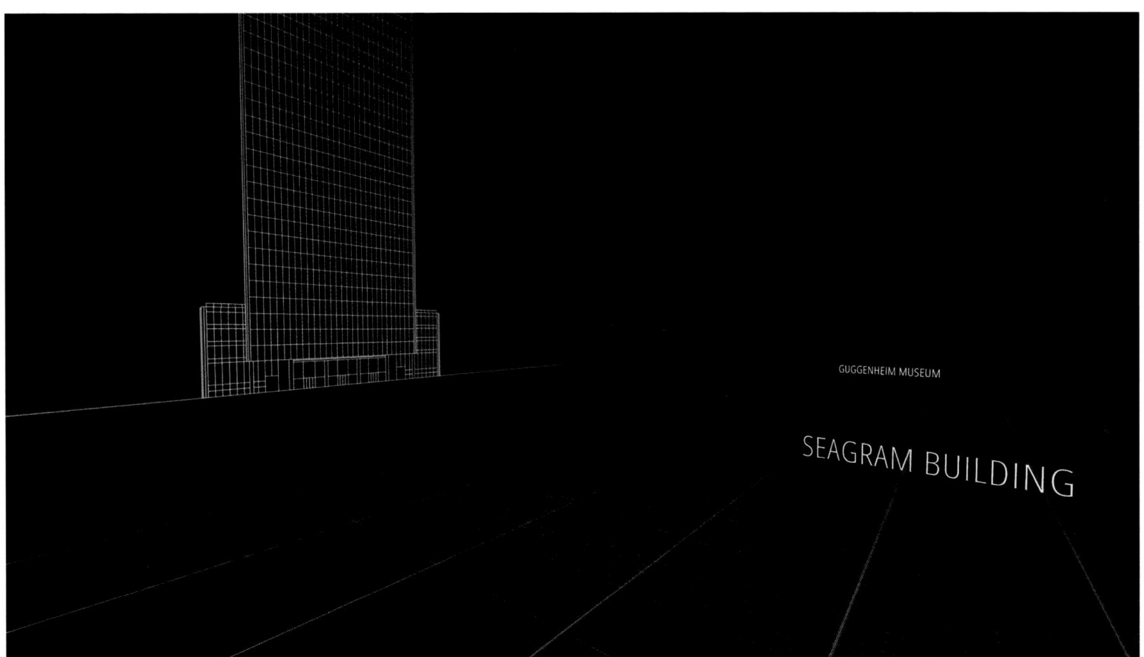

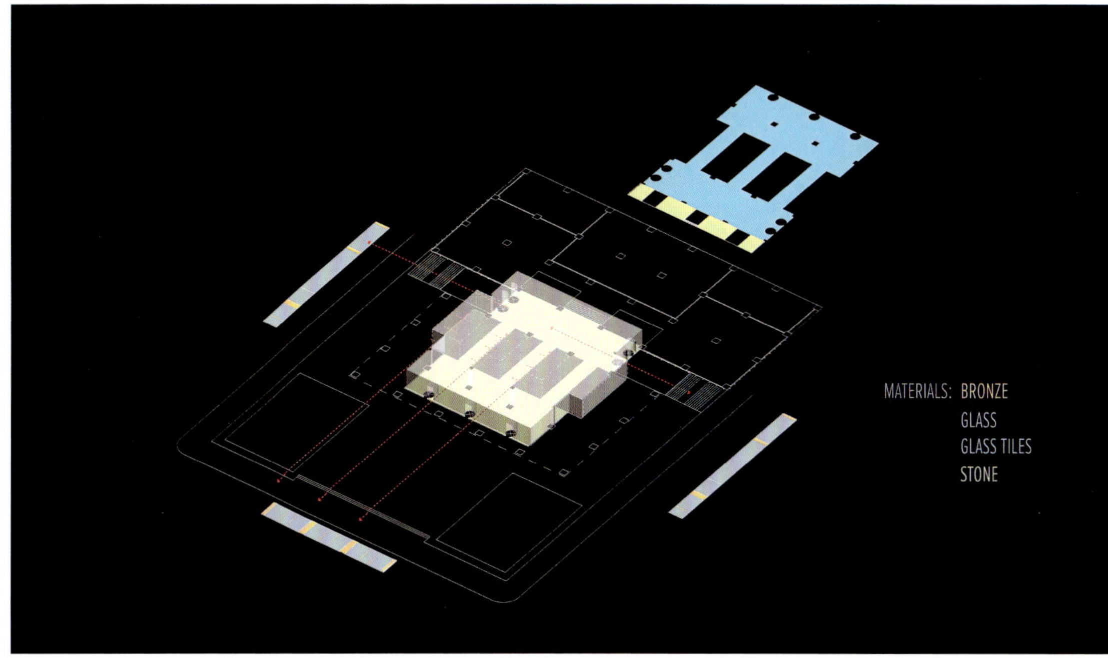

5-18 Digital street image of the Seagram Building, *Soundscape Architecture*, 2014. [Karen Van Lengen]

5-19 Photograph of the Seagram Building Lobby, New York City. [Karen Van Lengen]

5-20 Analytical diagram of the Seagram Building Lobby, showing volume, circulation, and materials, 2014. [Karen Van Lengen]

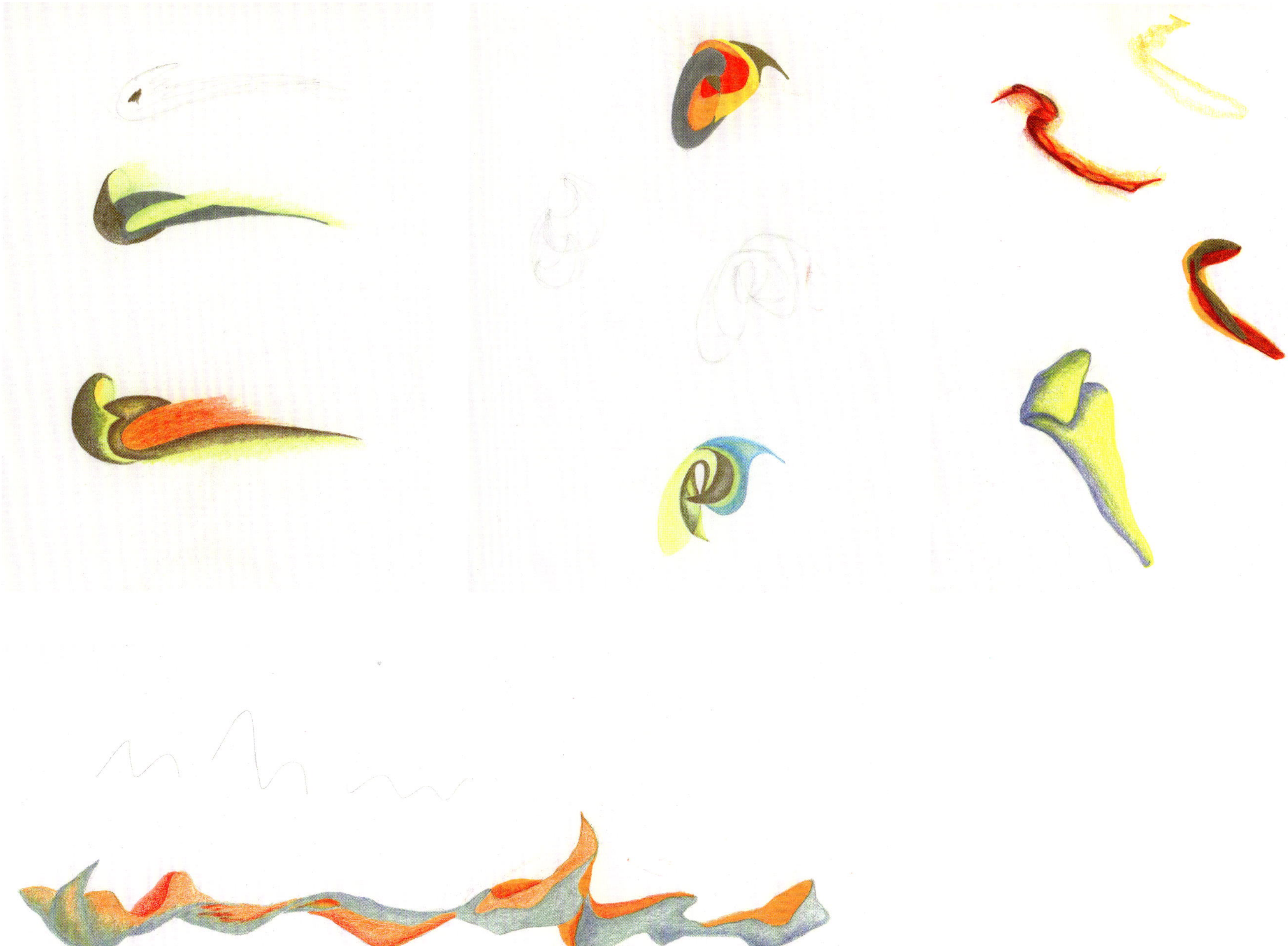

5-21 Drawings of the sounds of the Seagram Building Lobby, 2013. [Karen Van Lengen]

Guggenheim Museum

The Guggenheim Museum, designed by Frank Lloyd Wright and located at 89th Street and Fifth Avenue, is an original and masterful setting for the display of contemporary art exhibitions. Completed in 1959, the museum's concept includes a continuous circular ramp beginning at the lobby and circulating around the open interior hall, which culminates with a skylight. The ramp is the main circulation area for the viewing of art exhibitions, and it continuously overlooks the main space. In the course of viewing any exhibition, many people will move from the exhibition space on one side of the ramp to the edge of the main space on the other, and back again. This participation of the public in the central hall adds a significant sonic feature to any exhibitions located here. Of the five buildings presented in *Soundscape New York*, the acoustical condition here is the most varied, as the reactive sounds of the visitors will depend on the nature of the exhibition itself. For example, a James Turrell exhibition might evoke whispers of awe, as his renowned light installations often do. At the other end of the spectrum, the recordings made for this project were taken during the *Gutai: Splendid Playground* exhibition in 2013. The Gutai movement, founded by Japanese artist Yoshihara Jirō in 1954, explored experimental art whose "calls for vitality, play, and new artistic frontiers served as a jolt to a culture of consensus."[2] The exuberant artworks that filled the Museum's central hall conveyed a sense of joy and hope as one processed through the show. In creating the many recordings of the space during this exhibition, several included the screams of a young toddler. These were not screams of anger, frustration, or

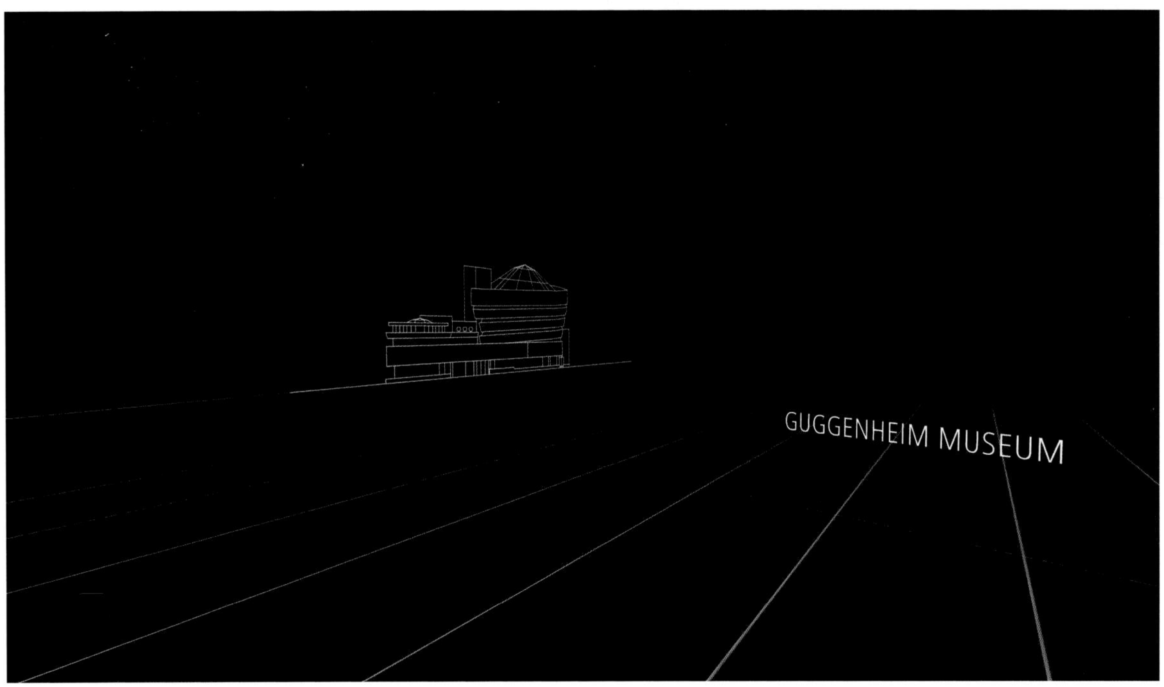

5-22 Digital street image of the Guggenheim Museum, *Soundscape Architecture*, 2014. [Karen Van Lengen]

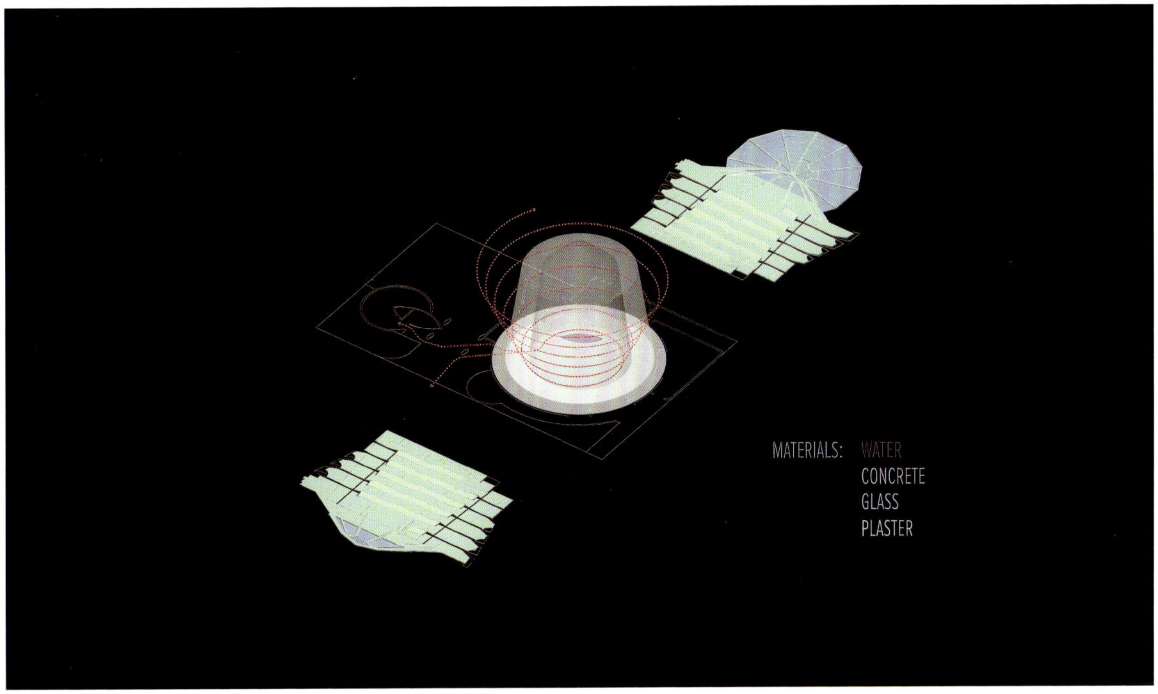

tiredness—but rather those of joy at encountering this great space with colors, sounds, and interesting objects that clearly delighted this child. In the words of Yoshihara Jirō:

> Gutai places an utmost premium on daring advance into the unknown world. Granted, our works have genuinely been mistaken for Dadaist gestures. And we certainly acknowledge the achievements of Dada. But unlike Dadaism, Gutai Art is the product that has arisen from the pursuit of possibilities. Gutai aspires to present exhibitions filled with vibrant spirit, in which an intense cry accompanies the discovery of the new life of matter.[3]

The Guggenheim's soundscape that day perfectly encapsulated these possibilities and the joyful cry of discovery.

5-23 Analytical diagram of the Guggenheim Museum atrium, showing volume, circulation, and materials, 2014. [Karen Van Lengen]

5-24 Photograph of *Gutai: Splendid Playground*, an exhibition at the Guggenheim Museum, New York City, 2013. [Alessandra Commons]

GUGGENHEIM 8/13

5-25 Drawings of the sounds of the Guggenheim Museum, 2013. [Karen Van Lengen]

Sonic Landscapes

The next two projects, *Inside Central Park* and *Openings*, present two iconic landscapes: Central Park in New York City, and the Academical Village at the University of Virginia. The first landscape is 30 times as large as the second. In addition, these landscapes are embedded with vastly different programs that lead to different types of visitors, and therefore different sounds. The intimacy of the Academical Village lies in contrast to the enormity and diversity of sounds of Central Park. These differences have sonic implications as explored below.

Inside Central Park

Central Park is the vast green oasis in the center of Manhattan, covering 843 acres and visited by 42 million people annually, or an average of 115,000 visitors per day. The park's designers, Frederick Law Olmsted and Calvert Vaux, originally named it "Greensward Plan."[4] This entirely constructed landscape was built over a 15-year period beginning in 1858. The breadth of sounds in this park comes from the diversity of activities and landscapes found there.

Unlike the interior acoustics of the building examples, Central Park does not have one main sonic aura. To capture its many facets, Van Lengen made hundreds of recordings of the park over a two-year period. She walked all of the paths, visiting and recording the individual places, such as the Wollman Rink, the Western Shore Boat Landing, and the Swedish Cottage Marionette Theatre. She also made a mapping grid of the park that referenced the surrounding Manhattan grid. For these walks, Van Lengen crossed from east to west and back again along straight lines every 10 blocks. Of course, there were obstacles such as the Reservoir, the Ramble, and other obstructions, so the walking lines were not always straight, but due to this walking method there was much to discover in the sectional diversity of the park. She selected many of these recordings to be inserted into a sectional diagram of the park and an accompanying plan now included in the *Soundscape Architecture* website (see QR code to access this sound map).[5] In order to develop a new

5-26 Multiple sections through Central Park showing recording locations, 2021. [Karen Van Lengen]

5-27 Plan of Central Park showing recording locations, 2021. [Karen Van Lengen]

5-28 *[following pages]* Digital paintings and drawings of the sounds of Central Park, 2024. [Jim Welty and Karen Van Lengen]

composition for the park animations and drawings, Van Lengen and Welty worked with musician and composer Matt Wyatt to create the final composition for this project. This piece conveys not the particularity of each sound or place in the park but bends into the expression of the Park's overall sonic spirit. The animations and the following drawings were born out of this new composition.

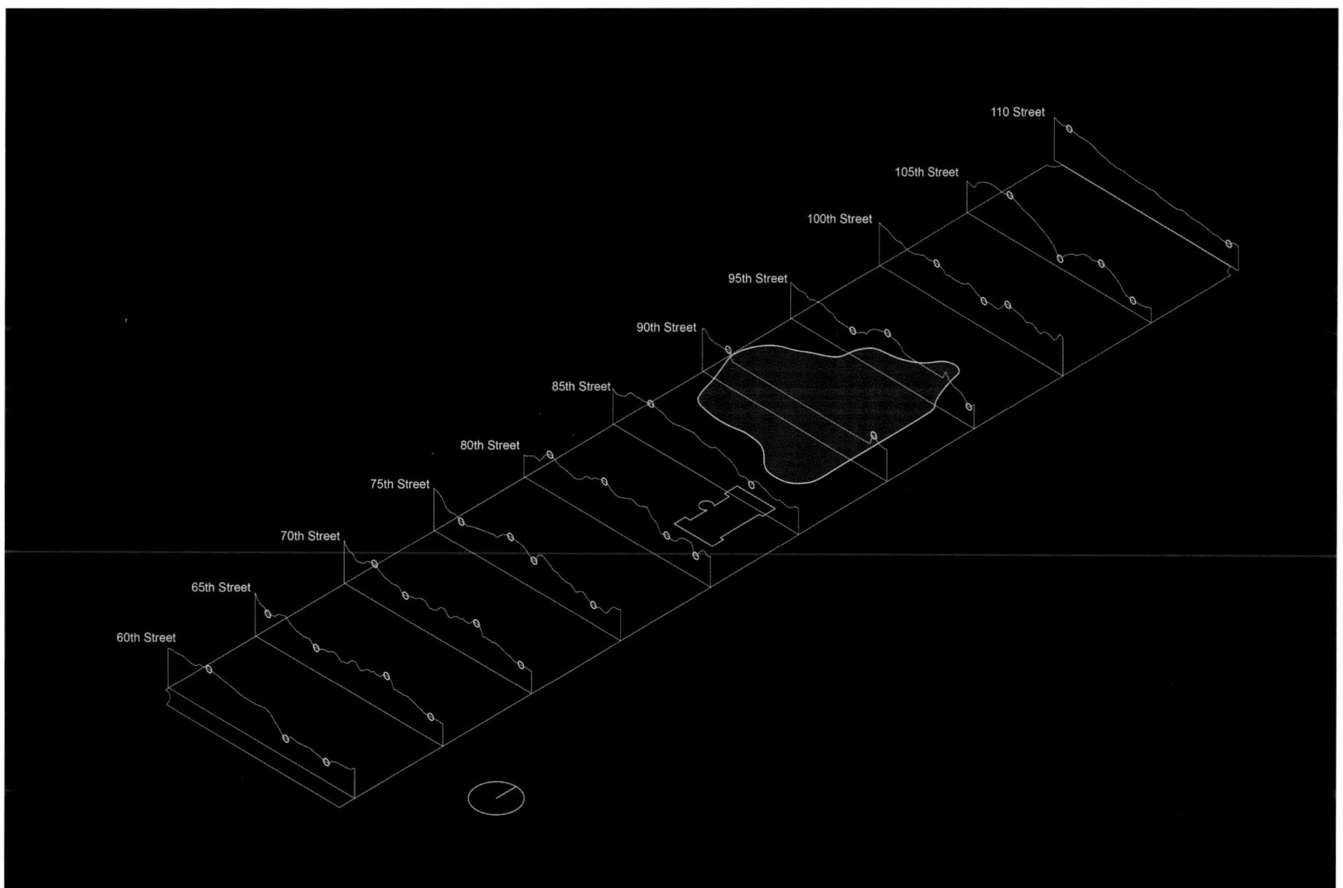

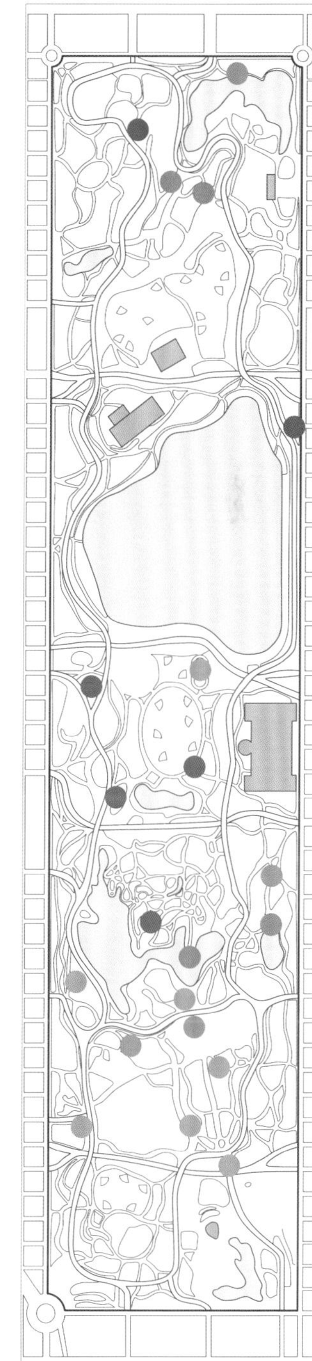

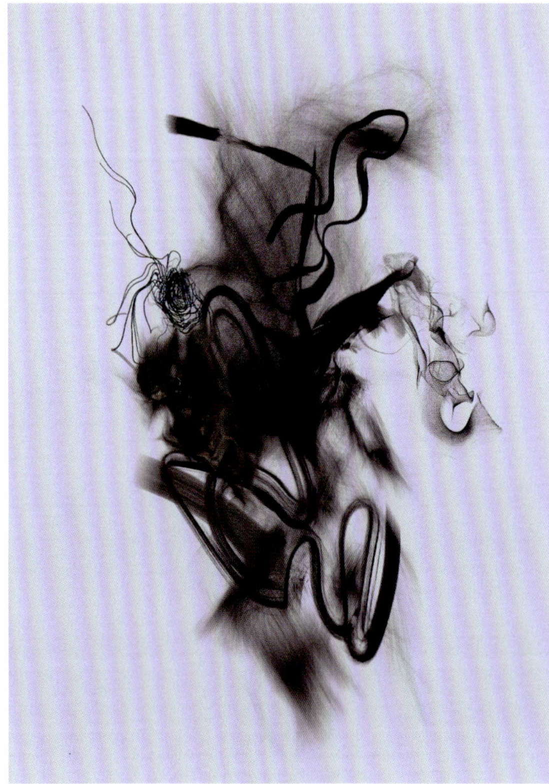

A HAHAHA

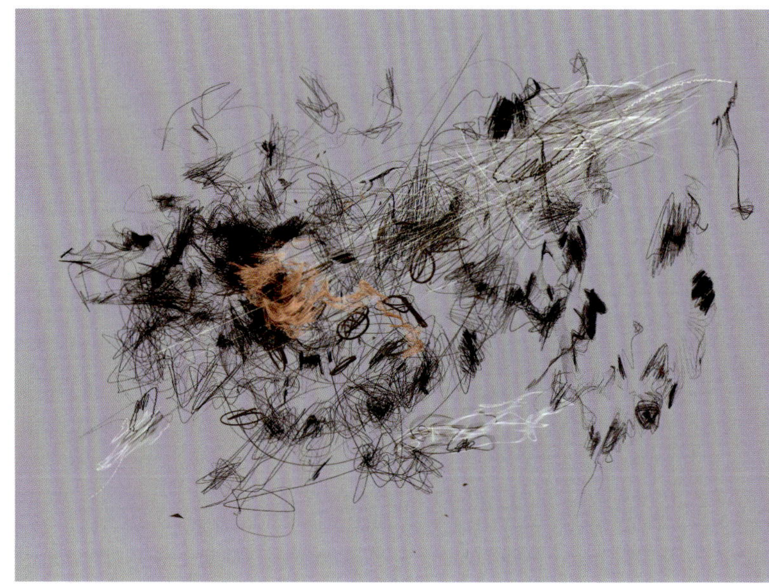

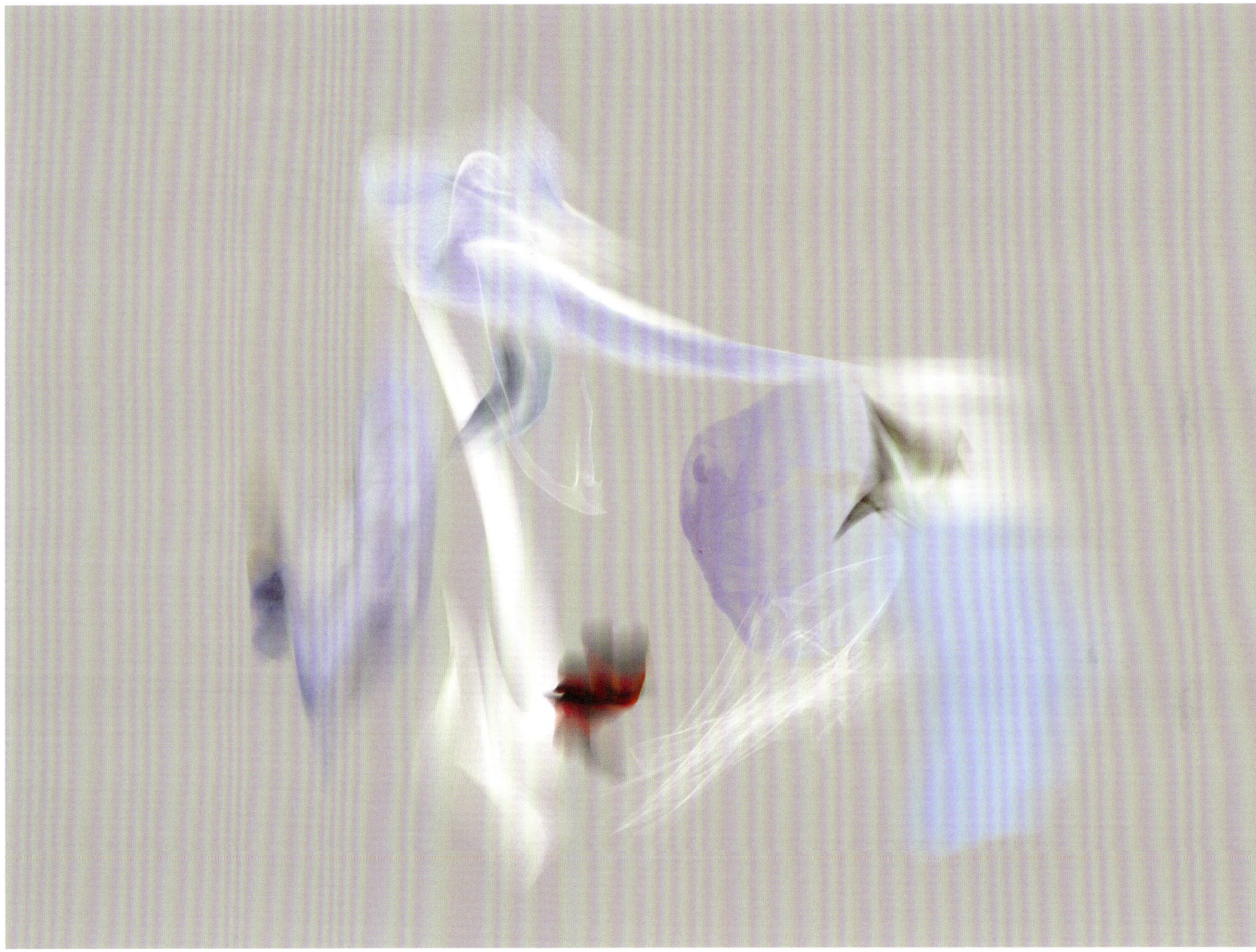

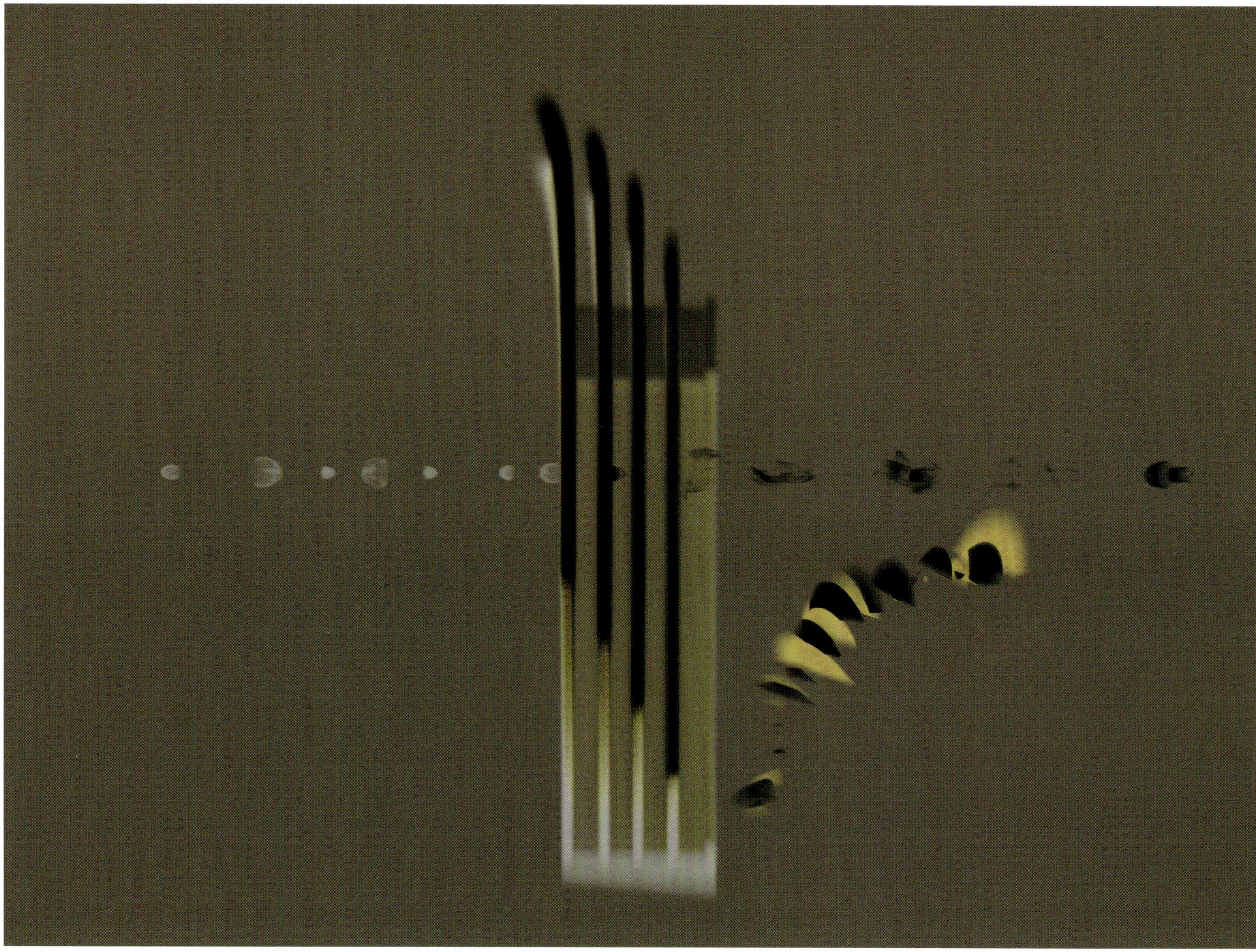

Openings

The Academical Village, designed by Thomas Jefferson and his consultants, was developed as the original "Grounds" of the University in Virginia that opened in 1819. The 28-acre site includes an inner grass lawn on three levels, framed by the Rotunda Library at its northern end and surrounded on its east and west sides with ten Pavilions distributed amongst student rooms. The inner complex is surrounded by a continuous arcade that serves as a unifying architectural element and also has special sonic characteristics. An outer ring of student rooms and ancillary buildings (called "hotels") encloses the gardens that lie between these outer and inner structures.

5-29 Aerial photograph of the Academical Village at the University of Virginia. [Woody Wingfield]

There is an intimacy inherent in the Academical Village due to its very human scale and academic programs. Although many significant texts have been written about the Academical Village over the course of its history, there is almost no reference to its sonic characteristics, which significantly reinforce a sense of human scale and occupancy. The largest open space inside the complex, called the "Lawn," is arranged on three levels descending from the Rotunda at its northern end to the lower-level lawn. The sounds of the Lawn vary, as they commemorate important University events such as Graduation and Convocation, as well as informal activities of student and faculty life. It is a sonically active space. For nearly 11 years, Van Lengen, Welty, and their daughter Kiri lived in the architecturally prominent Pavilion IX, during Van Lengen's deanship at the School of Architecture. Sound was ever-present. There were late-night student parties, outdoor movies, and classes. There were streakers, frisbee players, dogs, and alarm clocks going off at all hours in the student rooms. The Lawn at the Academical Village was a neighborhood of diverse sounds. In addition, the distinctive architectural arcade, which frames the Lawn, resonates with the voices, footsteps, and other sounds that travel along this enclosed space, depending on the time, the day, and the season. Sitting in the dining room of Pavilion IX, located directly adjacent to the colonnaded walkway, one could overhear every word and every footstep for a long length of the arcade. This intimacy, though sometimes disruptive, was a reminder that one shared this space with others in an ever-changing community of people and events. This specific sense of intimacy, unlike a more generalized urban soundscape, brings a unique sense of being inside this world of higher education and its mission to educate an American public.

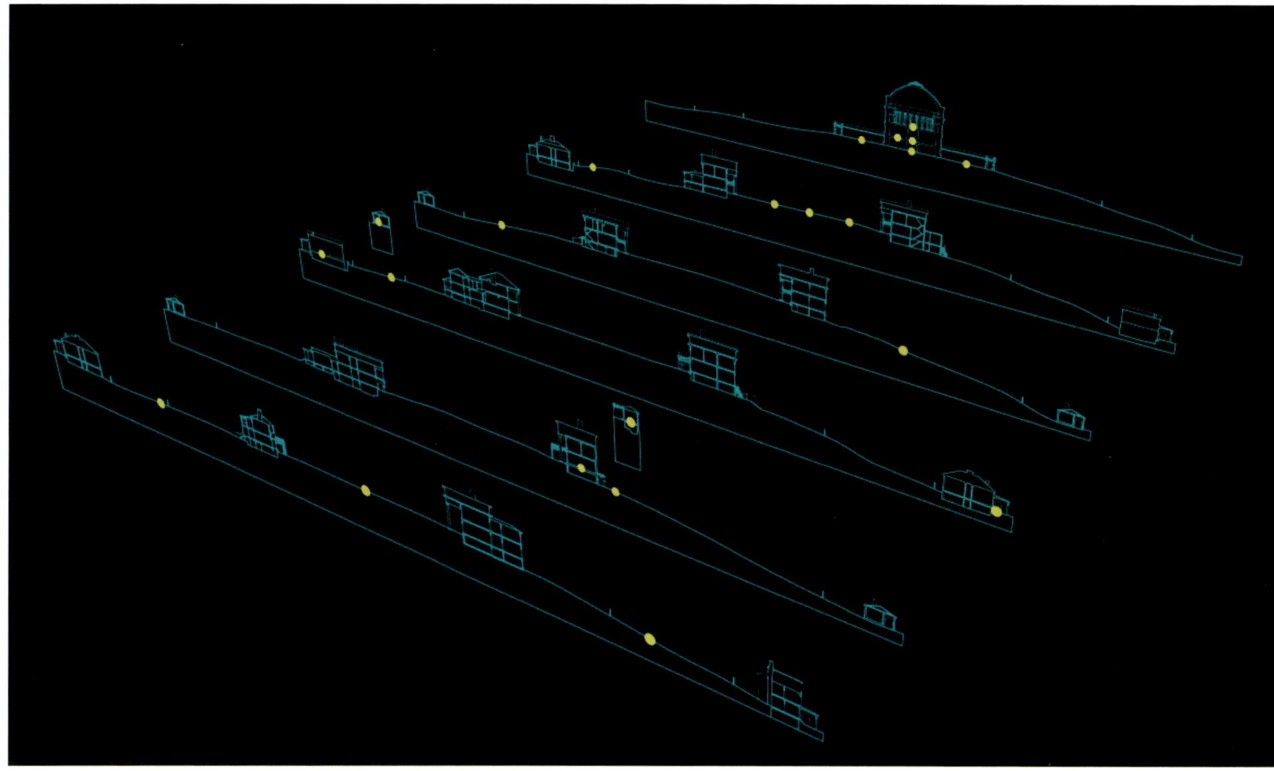

5-30 Sectional map of the Academical Village showing locations of recordings, 2015. [Karen Van Lengen]

During those same years, Van Lengen developed design programs and seminars that studied the sonics of the Academical Village. As part of their course work, the students recorded and drew their sonic impressions of the Academical Village. The website we developed, titled "Listening to the Lawn," includes proposals that represented their sonic memories of this World Heritage Site at the center of their university (see QR code to access this website).[6]

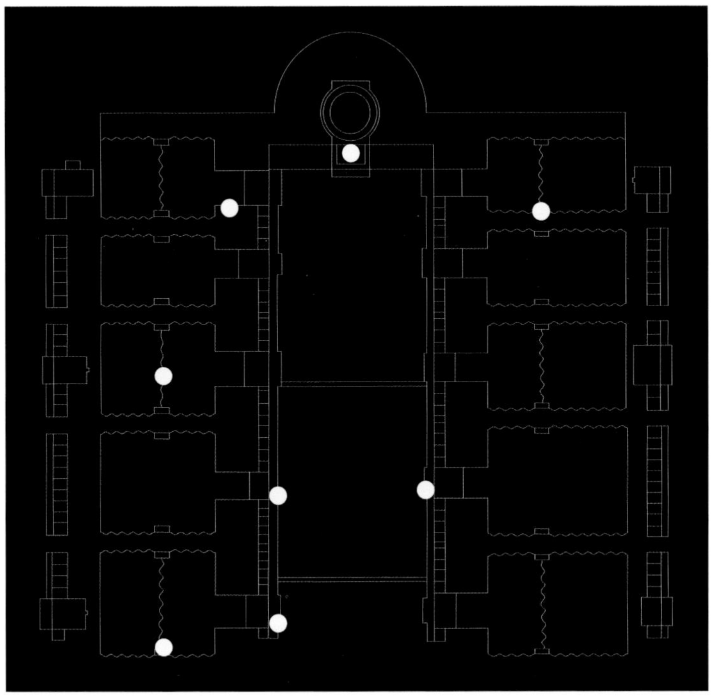

5-31 Plan view of the Academical Village showing locations of gates and doors used for the recordings, 2016. [Karen Van Lengen]

When Van Lengen and Welty later commenced their own listening project, *Openings*, they searched for characteristic sounds that would tie together their personal and public memories of this place. These sounds came from the gates and the doors. There are over 40 gates and 50 doors in and out of the various spaces and landscapes at the Academical Village. The gardens are open to the public daily, and the sounds of these garden gates are reminders of the public nature of this Village.

These gates are all sonically different. The clicks of the handles have different pitches and are accompanied by the diverse set of squeaks and vibrations from the door hardware. The humidity that swells the wooden gates brings different sticking sounds. Some gates rattle shut with a metal, self-closing apparatus. And then there are the doors, which open into public spaces such as the Rotunda, or the classrooms located in Pavilion VIII. They too are distinctive. The lower-level entrance to the Rotunda is a musical performance in four parts. As visitors enter this gateway it is announced to all, with a medley of clicks, cries in multiple pitches, vibrations, and bangs. And Pavilion IX, which, due to its curved wooden doors that complete the exedra form at the entrance, has a distinct click of the lock, and then a staccato opening of the two leaves of the door, followed by a heavy bang, which confirms the curved section has closed properly.

For the *Openings* project, Van Lengen and Welty chose four gates and four doors to represent this sonic wonderland. Welty's base drawing of the Academical Village is a visual sound field from which all of the gate and door drawings emerged in their animated forms, accompanied by their various "voices." Unlike the building interior projects mentioned above, the landscape here is not a single interior volume, so this open condition prompted Welty to view this animation as a field seen from the perspective of a bird flying overhead (see QR code to access the animation).

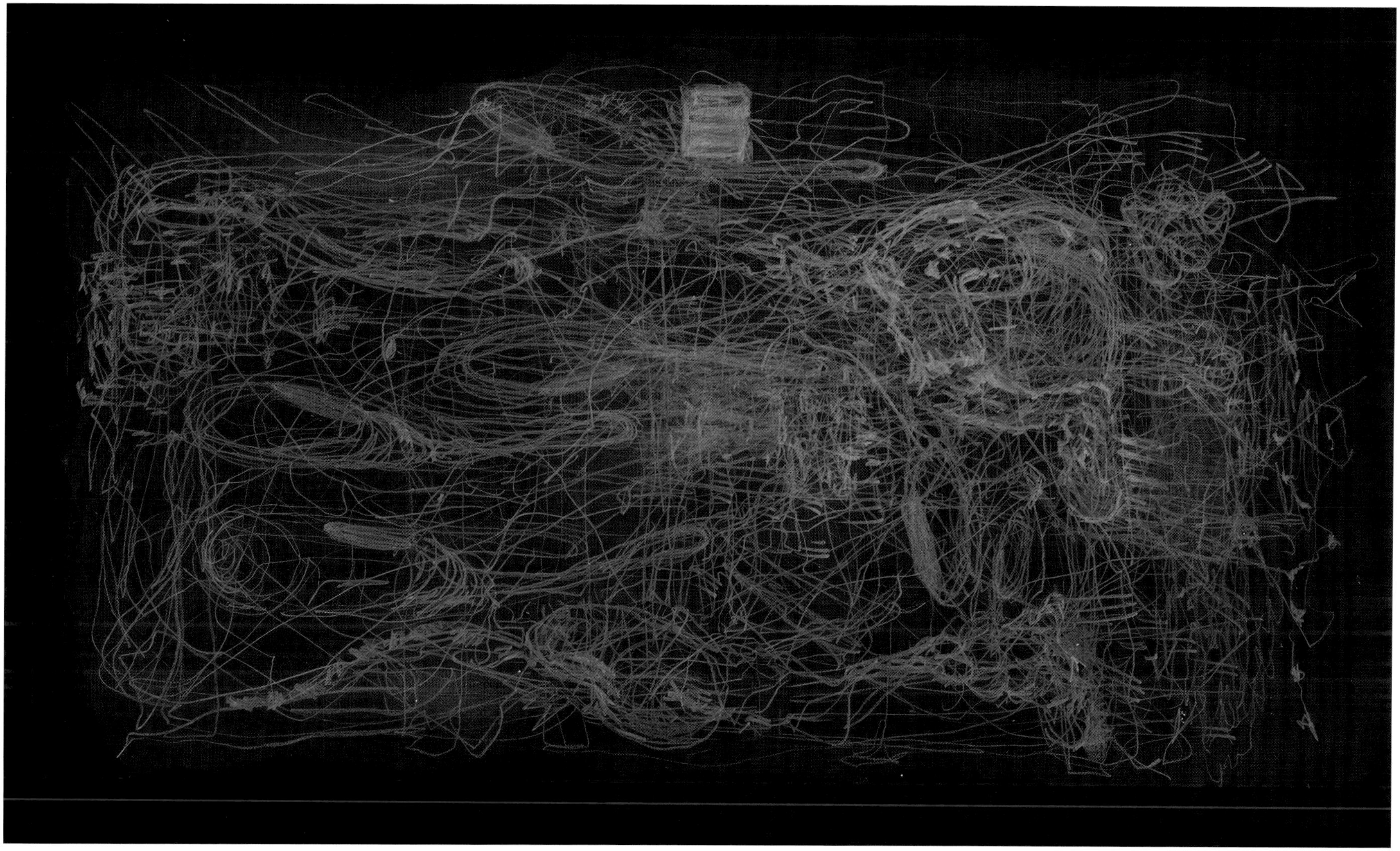

5-32 Drawing of the Academical Village atmosphere, 2020. [Jim Welty]

5-33 *[following pages]* Hand drawings transformed into digital drawings: Pavilion I (top left), Pavilion II (bottom left), Pavilion V (top right), Pavilion IX (bottom right), 2018. [Karen Van Lengen and Jim Welty]

10/29 PAV.I GARDT - FIRST HALF + 2ND HALF

2 43 6/11/19

Pavilion 5, Range Gate 2/13/19

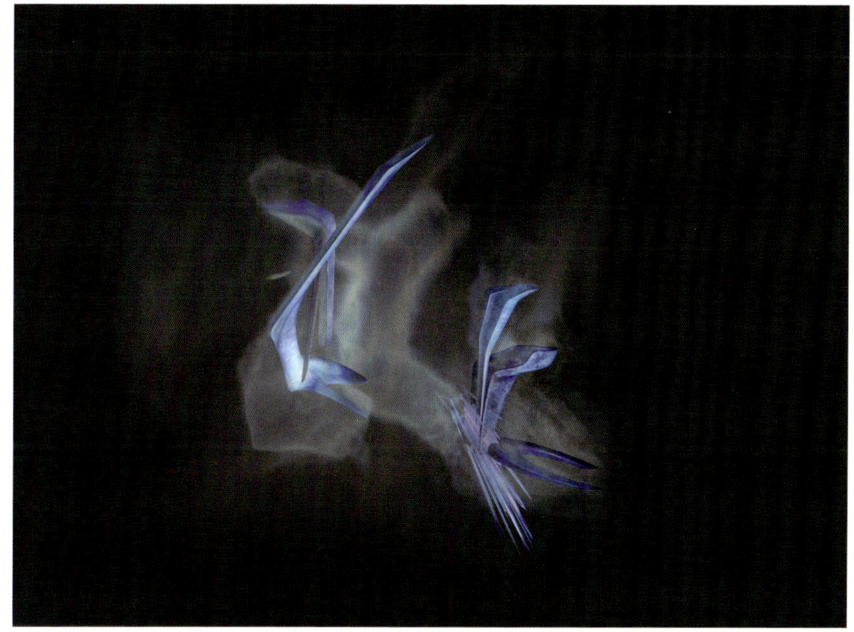

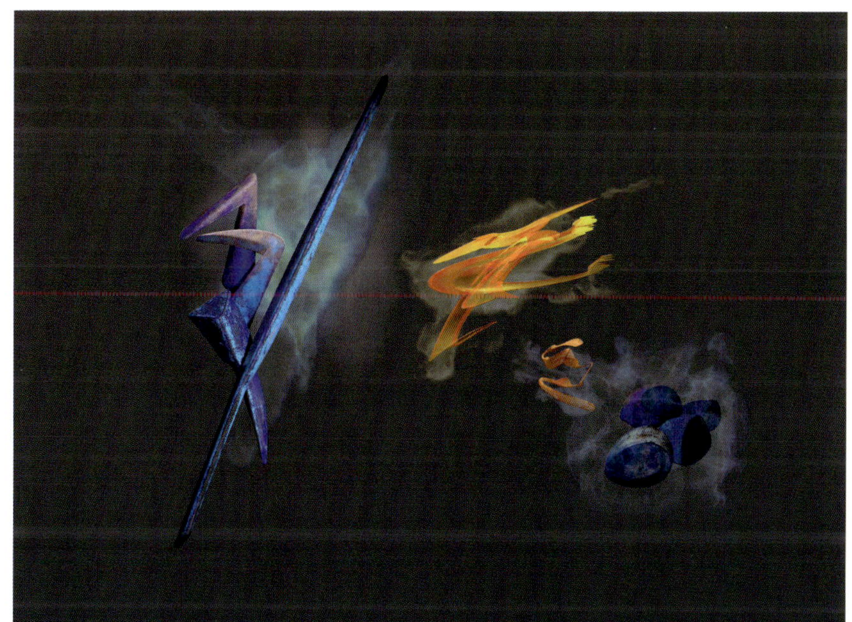

Gates to Gardens and Forking Paths into the Public Realm

Peter Waldman

> Narcissus, then, goes to the secret fountain in the depths of the woods. Only there does he feel that he is naturally doubled. He stretches out his arms, thrusts his hands down toward his own image, speaks to his own voice. Echo is not a distant nymph. She lives in the basin of the fountain. Echo is always with Narcissus. She is he. She has his voice. She has his face. He does not hear her in a loud shout. He hears her in a murmur, like the murmur of his seductive seducer's voice. In the presence of water, Narcissus receives the revelation of his identity and of his duality; of his double powers, virile and feminine; and, above all, the revelation of his reality and his ideality.[7]
>
> —Gaston Bachelard, *Water and Dreams: On the Imagination of Water*

Karen Van Lengen insists on being present in the world across scales through the resonant and aqueous depth of soundscapes. In her 2022 essay "Lacrimae Rerum, Tears of Memory," I was viscerally re-engaged, as a former swimmer myself, with Van Lengen's long-term immersion in the sounds of fluidity, subliminally recalling the haunting power of uterine buoyancy. In the spirit of fluidity, Michael Graves would remind me, in my apprenticeship with him, that spatial ideas were in the air, a communal primal atmosphere we all share in a world without the preconditions of orientation and gravity. It is dark down there, but the sounds of amniotic fluids echo at the pace of a heartbeat. This specific essay, "Lacrimae Rerum," alerts us to tune into the soundscapes of three recent urban projects found in public oases of resilience and renewal in Paris, France; Kassel, Germany; and a remote site in Iceland.

Reading this "Lacrimae Rerum" essay early on, I was sparked by Van Lengen's early valorization of the fluidity of memory, of the resonance of histories. This initiated a catalytic recall I encountered in 1986, to my surprise in the aqueous conditions of the Dallas Institute's translation of Bachelard's thesis on Sound and Memory, resonant as audible

Echo serving to complement the visually self-reflective Narcissus. Bachelard, in *Water and Dreams*, makes the case for sound as the requisite counterpart to the visual.

Decades apart, Karen Van Lengen, Jim Welty, and I have all been infected by our foundational years of cacophony in New York City before discovering Jefferson's Garden. We share lingering appreciations of the John Cage noises of great spaces, offered to us to appreciate the public realm—where Politics is understood as the Art of the City, a shared public realm. We also share *here and now* the great Lessons of the Lawn as a spatial and audible ideogram of Jefferson's Dream of a Vast open-ended Trans-Continental Imagination.

Van Lengen and Welty, in their collaborative work, identify two distinct scales for me to appreciate anew: one, the modest scale of the Gates and Doors to the Pavilions and Gardens of Jefferson's Academical Village, and the second, the monumental scale of five great public rooms in New York City. Each of the Gardens are as distinct as the corresponding Pavilions, and so are the Gates and Doors, as instruments, a symphony of distinct sounds connecting wood to wrought-iron hinges embedded into the mortar joints of Virginia Red Clay fired brick Serpentine walls in these Edenic Gardens. Each garden has multiple Gates, some to enter and others to exit, by tradition open to all from sunrise to sunset. Jefferson's Gates are never singular in material composition, as a wrought-iron gate might be, but always a hybrid, collaborative effort of a master carpenter, a skilled blacksmith, and a master mason whose frictional assemblages yield the resultant sounds Van Lengen and Welty record in sound and light.

The soundscapes of the great New York City public rooms we remember are re-presented in the Reading Hall of the New York Public Library, the previous site, significantly, of the first New York Reservoir on Fifth Avenue in the mid-19th century. Water and Dreams and the Role of Echo are recurrent leitmotifs here in this first Public Soundscape. Van Lengen makes the case that the sound generated from the movement of a heavy wooden chair across the terra-cotta tile floor moved the reader from the immediacy of the written page to an awareness that one is not alone, but in a great hall accommodating the generosity of echoes of inanimate furniture, and in the company of Citizens and Strangers. She makes the connection that the Two Stone Lions out front, Fortitude and Patience, have transposed their roars into the palomino-hued, massive chairs within, in a call and response, gospel-like sequence of echoes. The whisper of the mechanical book retrieval system adds another dimension of white noise, which meters one's progress, page by page and chapter by chapter. Here, crafted wood, heavy metal, and massive stone coffers provide base conditions to the patter of heels, and the flipping of pages made by the reader.

The next great hall is Grand Central Terminal, not in stationary silence as was the Public Library but filled with a cacophony of moving pedestrians ebbing and flowing, the metronome of the Great Clock, the multitudes of announcements, a symphony indeed. On a lower level is the sound of the shucking of oysters, with counterparts of crisp wrapping of both light baguettes and the ponderous sacks of rye and pumpernickel. Grand Central Terminal hums with the constant roars of departing trains going north, as morning commuters register the staccato scales of shoes both ascending stone steps leading to the south-facing monumental portal and descending into the subterranean subway system, which offers a counterpart as tympanum, as percussion, in response to the metallic screeches of departing trains. For those with time on their hands, one might look beyond the ticking clock that serves as a metronome to the celestial soffit of golden constellations depicted as dynamic figures read against this field of systems, within systems modeled after the great bath halls of Antiquity. Water and Dreams and the role of Echo also haunt this great symphonic space.

Michael Benedikt, author of *The Secret Lives of Buildings*, might be animated by these first two examples of Van Lengen's soundscape analysis. However, the next two great rooms are another hybrid spatial type, the lobbies leading to places of work beyond, transitions from the public street to the private office. Rockefeller Center and the Seagram Building, constructed three decades apart and funded not as public infrastructure serving both Citizens and Strangers but by private corporate entities, are also aligned almost exactly to the west and east of Fifth Avenue, and are attuned to the hybrid ranges of the Art Deco Gershwin era; the Seagram Building then echoing the understated metering of a reductive Philip Glass counter-composition of the late International Style. Both are dominated by the monumentality of the assembled elevator shafts, Rockefeller adorned and dematerialized by the bas-relief of polychrome figurations, while the Seagram piers are unadorned, muted by the dynamic reflections of the whisky-toned tinted glass and metered by a Cartesian Forest of shining bronze mullions. Both buildings are set back unconventionally from the street alignment and do not articulate a frontal facade; rather both share systematic elevations and read more as towers than urban blocks.

These elevational gridded fields of frames serve as the staff line or staves in musical notation, establishing a meter against which notes or figures are read. Rockefeller Center has a facade transposed as a watery oasis of seasonal flora and gilded fauna on the ground as a forecourt, not just in plan but in section, culminating in the excavated grotto of the frozen ice-skating Rink—overlooked by a Gilded Apollo above—which echoes the sounds of skates and piped-in music. However, the Seagram Building has the dynamic soundscape of animated vertical water works, a muting moat of sorts that masks, in part, the rush of ceaseless automobile traffic of Park Avenue. It also makes a counter-landmark across the street from the sweat and showers of the New York Athletic Club, while masking the rattle of the Metro North trains departing from Grand Central Station a few blocks to the south. As the Academical Village is an ideogram of Jefferson's vision of a transcontinental nation, so Van Lengen's research on soundscapes has discovered a way to read the city's Cartesian fabric of the orthogonal street grid as the notational staff by which to read the monumental figures of a 20th-century city with the frictional mix of an architecture of democracy: a library and a propylaea confronting the emerging patronage of corporate capitalism in a pantheon and temple to would-be vestal virgins. This diagramming of urban mapping mirrors landscape architect Lawrence Halprin's notation system of cartography and choreography evidenced in his design for Charlottesville's Historic Downtown Mall whose Cartesian Forest is the counterpoint to Jefferson's democratic Lawn.

The first four Soundscape New York locations are sited in Midtown Manhattan distributed to the west and east of Fifth Avenue as a spine spanning north to south. The fifth and final soundscape of Van Lengen and Welty's quintet is the vast space of Frank Lloyd Wright's Guggenheim Museum again on Fifth Avenue, just 48 blocks to the north of the New York Public Library now facing Central Park and the Jacqueline Kennedy Onassis Reservoir. This soaring volume is a base drum whose concrete surfaces resonate with the whoosh of the elevators, the staccato of the pitter-patter of pedestrians spiraling up to heaven, a bit of a marketplace hall between gallery and gift shop, both a Tower of Babel and a Nautilus Shell looking into one's soul as if to follow this labyrinthine meander at the pace of one's heartbeat.

One might think Karen Van Lengen and Jim Welty have taken us through the requisite five great soundscapes of what is known as the "New York Five."[8] However, they have the appetite to include Central Park, right outside the Guggenheim stairway to heaven. The Guggenheim faces west and sees itself reflected in the Reservoir, now echoing with the giggles of children, the clopping of horse hooves, the concerts on Sheep's Meadow and Strawberry Fields. Further south, one engages the Central Park Zoo, with the roar of the lions, the splashes of the seals, the chatter of the monkeys, before arriving at the horse-drawn carriages of Central Park South and the Memorial Fountain at the Plaza. We are finally taken back to that great hall reading room just 10 blocks to the south, aligned just off Fifth Avenue.

Van Lengen and Welty commence with the metaphor or mnemonic device of a modest Garden Gate behind the scenes in the Piedmont Condition, and akin to Aeneas, they lead us through the Five Great sonic spaces in a huge cosmopolitan setting. But Central Park is, conveniently, the Sixth, and implies the Genesis cycle which ends in the Garden of Eden.

I will let the late Joseph Rykwert have the last word when he reminds us that the Origin of Architecture is initiated in the human spatial practice as Dance in *The Dancing Column*.[9] And I cannot imagine dancing without human voices and instrumental agencies as an archaic practice, to which retorts Sanford Kwinter in African Genesis, "what can be more Modern than the Archaic."[10]

Art

Everything in the world has a spirit which is released by its sound.[1]

OSKAR FISCHINGER

Van Lengen and Welty's Soundscapes: Process, Sound, and Space

Nana Last

Under the comprehensive moniker *Soundscape Architecture*, Karen Van Lengen and Jim Welty explore the intersections of sound, space, architecture, and the public realm. By visually reimagining the aural qualities of iconic architectural spaces and landscapes through the processes of recording, listening, editing, drawing, animating, and compiling, their resulting Soundscapes are as much a process as an outcome. Originating from visits to public spaces such as New York's Grand Central Terminal or the New York Public Library's Rose Reading Room, the collaborators' end products—animations and digital prints—are the outcome of a multi-stage interpretive and analytical process undertaken to reflect their memories of journeying through these places. Through visually staging that journey, Van Lengen and Welty endeavor to capture the sonic ambience of those spaces along with the more fleeting sound events that transpired during their visit.

In engaging the realm between sound and receiver, the Soundscapes encounter two historic sets of concerns. The first is a form of metaphysical questioning that can be referred to as the tree/sound problem; the second, the issues surrounding copyrighting the movement of human bodies in space. In this, the processes the Soundscapes embark upon can be seen to enter a fraught territory often defined by questions and oppositions, and emerge out the other side as gleeful art. Their success lies not in avoiding these concerns, but in embracing them. The glee that results at the far end of production—literally at times in the animations—is the very product of moving through that terrain.

What I am referring to as the tree/sound problem is a long-asked thought problem that has attained the status of an adage. It ponders: "If a tree falls in a forest and no one is there to hear it, does it make a sound?" The form of the question situates the observer within a seeming conundrum, one whose elusive compass imparts an ambivalence over the role of human perception in the world. Under this aegis, the predominate modes of response have typically rested upon differing definitions of sound. The first sees sound as vibrations that travel through the air or another medium and can be heard when they reach a person's or animal's ear; the second describes sound as continuous and regular vibrations as opposed to noise. The first definition associates sound with hearing or sensation, proposing that perception of air vibrations is a necessary component of sound, while the second approaches sound technically or mechanically as a variation of pressure that propagates through matter as a wave. In terms of the tree/sound problem, sound as mechanically understood will be deemed to have occurred if the tree falls and no one is there to hear it, but sound resting on the need for human or animal perception, will not.

Caught in this dichotomy, the tree/sound problem readily engenders responses that seemingly pit material state against sensation and perception. Further issues ensue from this division of sound as air waves versus perceived sensation. On one level, separating the technical or mechanical aspects of sound from its perception creates a split between a material and its properties, on a metaphysical level, the conundrum seemingly calling objective existence into question by asking us to consider if something can exist without being consciously perceived. The splits and potential new problems do not stop there but continue to create ever-new situations as technologies, capable of measuring the physical phenomena of sound by instruments other than ears, emerge to allow sound to be sensed outside of human or animal perception. Technology further complicates the situation in these cases by admitting the possibility that sound could be air vibrations registered by nonconscious entities.

The space the Soundscapes activate is exactly the gap between these various and widening definitions—between sound as wave and sound as perception, between material qualities, technologies, and human bodies. Their first move in engaging sound is to restage the problem to show that the space in which the sound occurs is a crucial component of this discussion. To this existing terrain they add a host of processes of listening, memory, and visualization, whose inclusion reimagines our relation to sound. The Soundscapes' ready occupation of this seemingly oppositional state makes clear it is less a problem to be solved than a wealth of material to engage with. Rather than see sound as a dichotomous conundrum, the Soundscapes show it to be a space imbued with the potential for innovation arising from the intersecting of people and memory, sound and vision, technology and the materiality of place.

The work of the Soundscapes begins with the tacit acceptance of an inverse relation between the precarity of sound and its important role in public space. From there, it sets itself the problem of how to show sound. How to transform the conundrum of sound as airwaves or perception to one of how to show—and share—sound's very connection, its ability, despite its ephemerality, to have lasting impacts. In this way, Van Lengen and Welty expand the premise of the thought problem to include the space of experience to which they happily admit themselves and invite a host of processes and technologies to follow. It is as though they took the dictum that sound waves need a medium by which to be transmitted and ran with it.

In creating the Soundscapes, Van Lengen and Welty instantiate a collective process that spools together sound waves, perception, individuals, memory, space, events, and multiple technologies into an array of moving and still visual worlds. Van Lengen describes the process of constructing the Soundscapes as advancing through a series of stages that begin with listening and recording, move on to editing, followed by drawing, animating, and reconstructing digital images. The Soundscapes are formed by the coming together of two main components: the animations, which instantiate the overall ambience of the place, and the sound object drawings, which relay notable sounds heard during their visit. While both sonic ambience and sound objects are generated from the same visit, each initially develops through its own processes then dovetails into the animations.

The process is initiated by listening and recording at the site. The premise of this stage of the process is that a place's overall sound ambience comes through in all of the recordings. To better understand and elicit that ambience, they typically record the space at different times of the day to capture the variety of sound conditions indicative of the place. The resulting edit conveys both the overall ambience of the sonic space and captures distinct sound events that transpire within it. Once the recordings are made, Van Lengen and Welty listen to them repeatedly with the goal of identifying the recording most consistent with their own memory of the space. Van Lengen then does two things, she edits the selected recording into a 60-second piece, and she identifies notable distinct sounds that happen within it; these are designated "sound objects." The act of naming, itself, initiates sound's transformation from more ephemeral state to visualized entity. The sound objects, which include such things as the tinkling of glasses from Harry's Bar in Grand Central Terminal or the trains arriving on its main concourse, are visualized in the sound object drawings.

While based on the 60-second edit, the process by which temporal sound becomes visual image is a product of individual human perception and recall. Van Lengen acknowledges this, declaring: "Drawing the sound objects is deeply personal. I listen to the sound fragment over and over until sound loses its identity of origination and becomes pure sound. It takes a while—and then I start drawing. I draw that sound many times, listening to the sound while drawing. Then—finally—I turn off the sound and draw what I remember, and those drawings become the final 'sound object' drawings." The acknowledgement of individual memory is important in this, as the individuals listening are as much part of the sonic experience of the place as every other aspect. Van Lengen repeats this process, drawing each of the sound objects multiple times until she feels the drawings to be "reminiscent" of that sound. Until, that is, the visual form of the transcribed sound object accords with her memory of it. Her goal in this process is not to focus on the source of the sound, but "to get at the pure sound" rather than "its generator." The process hinges on a transformation from sound as conveying its origins, to becoming solely the perception of the aural qualities of the sound waves without thinking "glasses tinkling" or "tree falling," as though there were no recognizable origins at all.

In taking this stance, Van Lengen situates this phase of their work—the process of drawing—at the mid-point between the two dichotomous stances, between sound creation from a source and the perception of that sound: between sound's emanation and its sensation by the human subject. It is at once the transmission of a particular event and yet dislocated from those origins to be something of its own. This state is the basis of the art that follows. It is the foundation for the transcription of sound into the visual realm—a process that transforms it from its ephemeral state to being fixed on paper, following which it will enter the animations and be returned to the public through the act of exhibition. Transmitted in this process is the memory of the experience of the space, a memory that harkens back to that origin, but is not exactly put back in the place from which it came.

The animations, as the next stage of the process, can be seen to do three main things. They create an overall visual structure that acts as the visual equivalent of the ambient sound space; they serve as the repository for the sound object drawings; and they animate the sound objects drawings. This stage sees a shift both in the role of the collaborators and the mode of the source material. While the sound object drawings become part of the animations, they are not the source of them. Although born from the same sonic space, such as Grand Central Terminal, the animations have their own unique history into which the sound object drawings are then situated.

This animation stage adds both additional source material and a new technology, that of the computer. Welty constructs the animations by overlaying photographs and other images of the originating site to build up layers of information. Constructed this way, the overlaying and embedding of multiple images of the space itself is designed to surpass representation of the place and become a purely visual structure. That structure is the visual correlate to the pure sound, which the sound object drawings transcribe. If the sound object drawings transpose sound events from the site and transform them into visual entities in the space of a drawing, the animations transform visual material from the site into a visual, temporal framework for the sound objects to come to life. In this way, they serve as the site for the sound object drawings in the same way the building interior served as the site for the sound events that occur within it. Once the animations are formed, the sound object drawings are brought into them, with Welty orchestrating their movements to coordinate with the place they occur on the soundtrack. Van Lengen and Welty describe this part of the process as constructing a "theater" of the ambient overall sounds within which the sound object drawings appear as characters "playing their sounds in an ever-changing theatre stage."

The digital art prints constitute the final stage of the process. Although they are generated from the animations, the digital prints are not isolated moments from it. Rather they, similarly to the animations, also overlay information by gathering multiple moments from the animations and reconstructing them into a single image that captures distinct views or memories of the sound objects. Together the entire process—including its various technologies and mediations—creates an assembly of innovative tools and methods that transmit the experience of a place and time through the framework of sound. Testifying to sound's manifold modes of being, this process, which exhibition extends to the viewer, follows the path of sound through the individual, and memory, and emerges as visual representation—drawings, animations, and digital prints. What makes it come to life are its processes of transcription.

On one level, we can clearly see the processes generating the soundscapes are those of recording, listening, editing, drawing, and animating. On another level, the processes can be thought of as a series of spooled transcriptions. Each layer of the process spools information from the previous stages into a new medium that constitutes the next stage. Through this process, the series of transcriptions situate the work in the realm between those who heard and experienced the sound and the public who will see the exhibited work. They tie the recording of the physical space of sound—through memory and visualization—to another public. Paralleling the action of sound waves through a medium as a form of transmission, the ability to transmit information across these stages resonates with a second issue the soundscapes invoke, that of the challenges surrounding copyright in the arts.

If the tree/sound problem sets up a territory of inquiry the soundscapes pass through, the issues surrounding copyright of another ephemeral event, that of dance, lays the ground for understanding their method of response. The 20th century saw the realm of dance and choreography play a unique role in the creation of copyright law. Prior to 1952, when, for the first time, a copyright was awarded to a choreographed dance, no basis had been established upon which to copyright such a dance. This changed when Hanya Holm successfully registered a copyright for a dance. Success in this precedent-setting case rested on two things. The first of these was the understanding of choreography as a specified sequence of steps and movements as opposed to a collection of isolated moves. The second part of the copyright success rested on its ability to materially and legibly convey this information. The innovation in Holm's case lay in conveying that information in a written notation system for recording human movement, in that case through a Labanotation score, as part of the copyright application.

We can see the sound object drawings engaging both these aspects—sequence and notation. More precisely, the drawings are transcriptions that visualize the selected sequence of sounds Van Lengen recognizes as sound objects. Notations and transcriptions are related terms. Transcriptions can be thought of as the broader category, designating written or printed representations of something. Notations are more systematic, defined as the act, process, method, or an instance of representing by a system or set of marks, signs, figures, or characters as developed in art, science, mathematics, and logic. Both notations and transcriptions perform their function of transmitting information by working through another media or mode of being other than the original one. This is what allows them to convey information from one realm to another. As the first stage in the process to visualize sound, the sound object drawings are a transcription that serves as the hinge between the originating sound and its presence in the animations. They are the critical stage in moving from a more ephemeral realm through acts of memory, and on to a public realm of visual/animation/exhibition.

The Soundscapes, while similarly acting to convey information, do not propose a systematic one-to-one correlation of visual mark to sound wave that aspires to be a fixed legal notation. Rather than be a direct notation, the Soundscapes' transcriptions are descriptive. They capture the transmission of air waves to the perception of them. Interface becomes art. They register the cumulative process of sound within the social space of the building and as recalled in memory. As copyrighting a dance creates a site in which bodies/flesh in motion meet with legal codification, the Soundscapes similarly relay configurations of humans and technology. Transcription here is responsive, not preset. Unlike action in a legal realm, they act in an interpretive one. Yet, while not the kind of notation that may be used in a legal setting, it is interesting to note that legally not all aspects of copyrighting are preset either. The courts, for example, do not set out in advance exactly where or when a single unit turns into a sequence, so perhaps the two are not as far off as first appears.

As sequences happen over time, visualizing a sound object entails capturing that sequence of sound waves, both in its aural qualities and duration. The drawings do this by creating an interface between the ephemeral event and its transmission to new situations—places and times, including their exhibition. In that way they are transcriptions that transmit information. Unlike the initial recordings of the place that directly transmit the sound to a fixed state, the drawings, animations, and digital art prints inscribe the individual experience of that place. That is what they recall and record, and in so doing, they inscribe the interface of bodies as receptors of the sound with the public space itself. Their transmission of information and experience, however, is transformative in yet another way. As Van Lengen and Welty's undertaking ties sound sequence and visual transcription across various media, the techniques they introduce do not end there. The connections they establish and convey extend into any future experience of these places, making us not just more attentive to the sounds we may encounter there, but able to visualize them as well.

Digital Paintings of *Soundscape New York* and *Sonic Landscapes*

Jim Welty

The digital paintings of *Soundscape New York* and the Academical Village that follow were initially created for our 2020 exhibition at the Branch Museum in Richmond, Virginia. Since then, we have expanded this body of work to include paintings of Central Park. Together, these paintings represent the culmination of our journey to capture our perceptions and memories of these iconic spaces, by beginning with the act of attentive listening. Each painting is intricately linked to the process of listening, recording, composing, drawing, animating, and finally creating the digital paintings of these sonic engagements. These pieces not only signify the final chapter of our exploration but also evoke the transient atmospheres we have come to appreciate through the art of listening.

New York Public Library

Grand Central Terminal

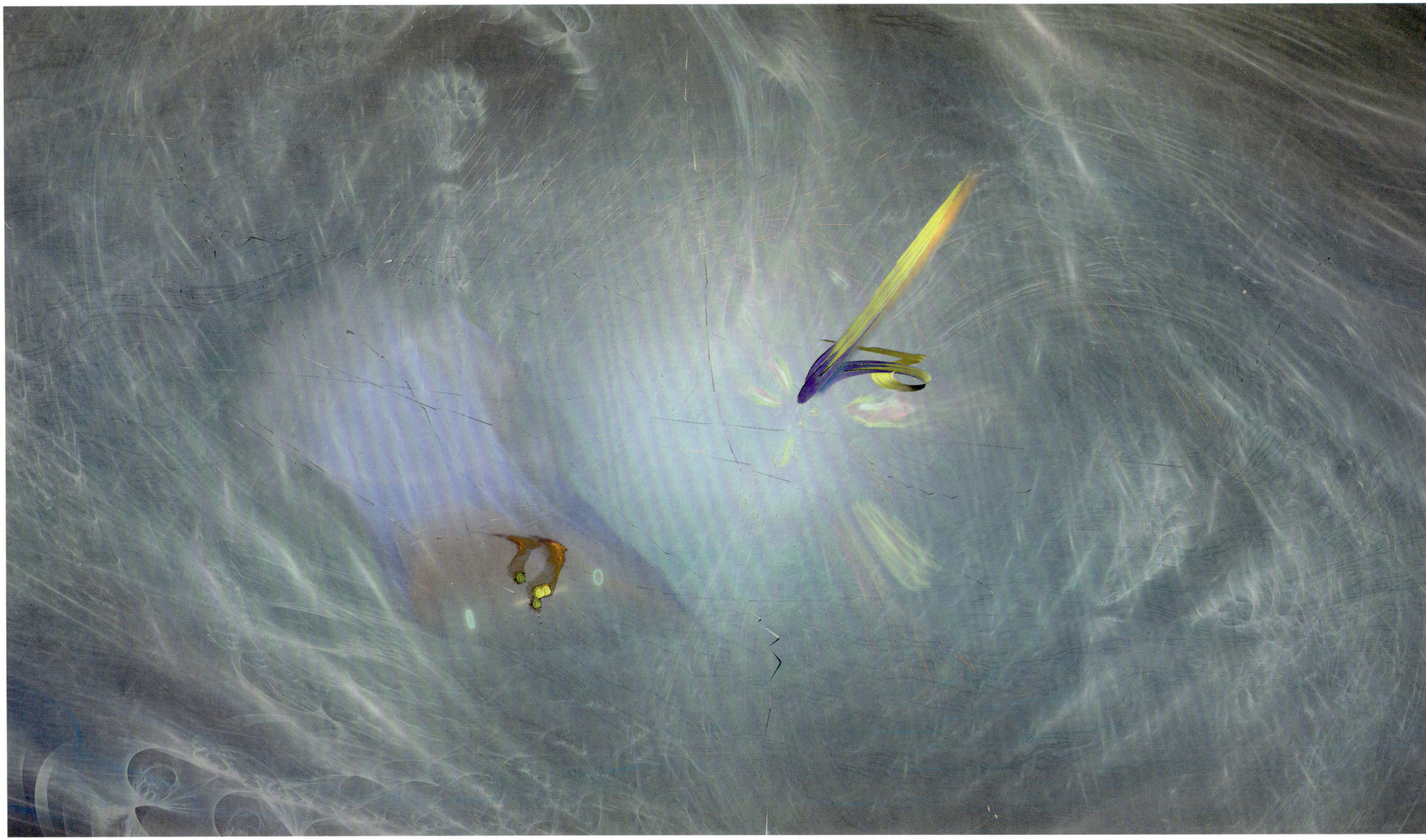

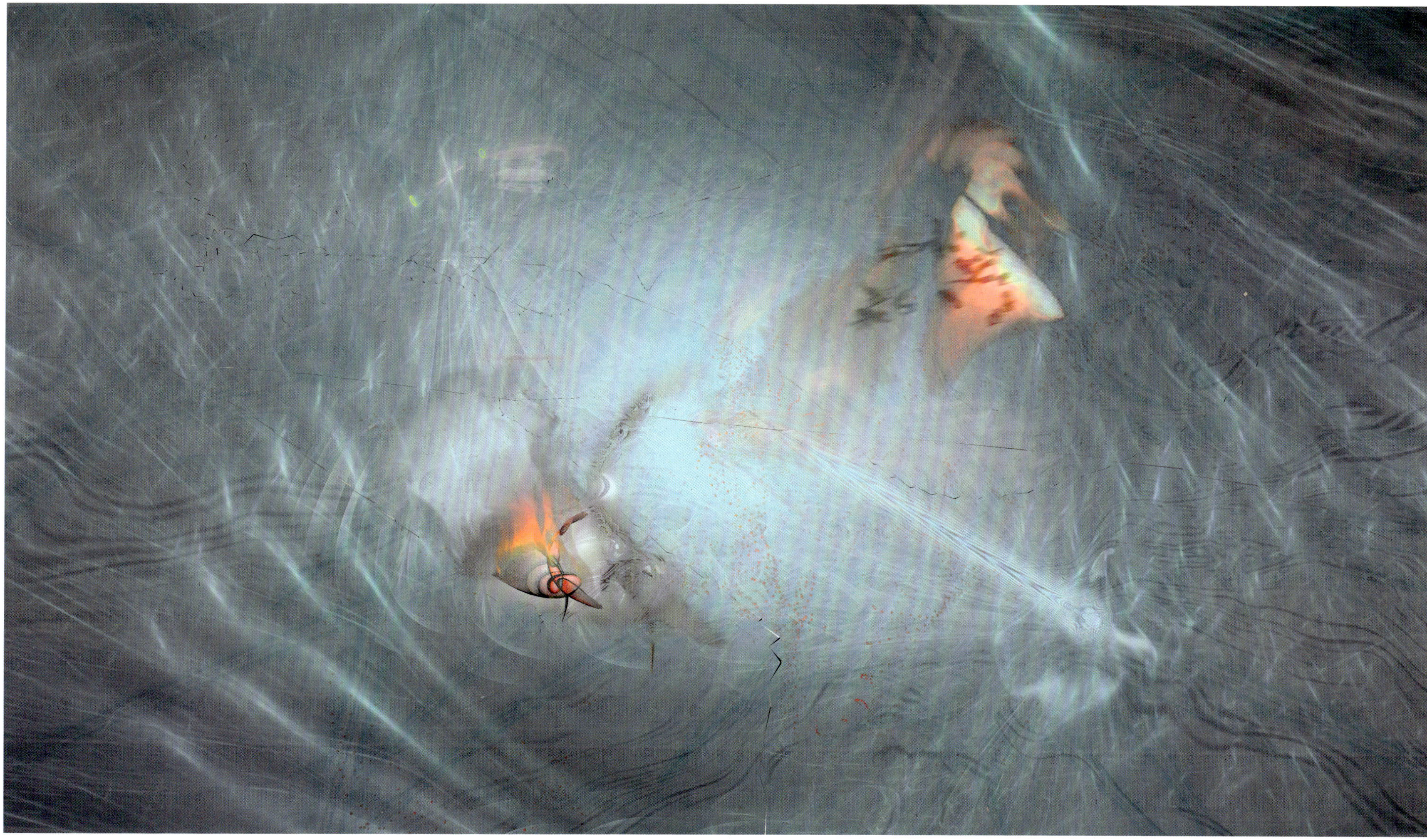

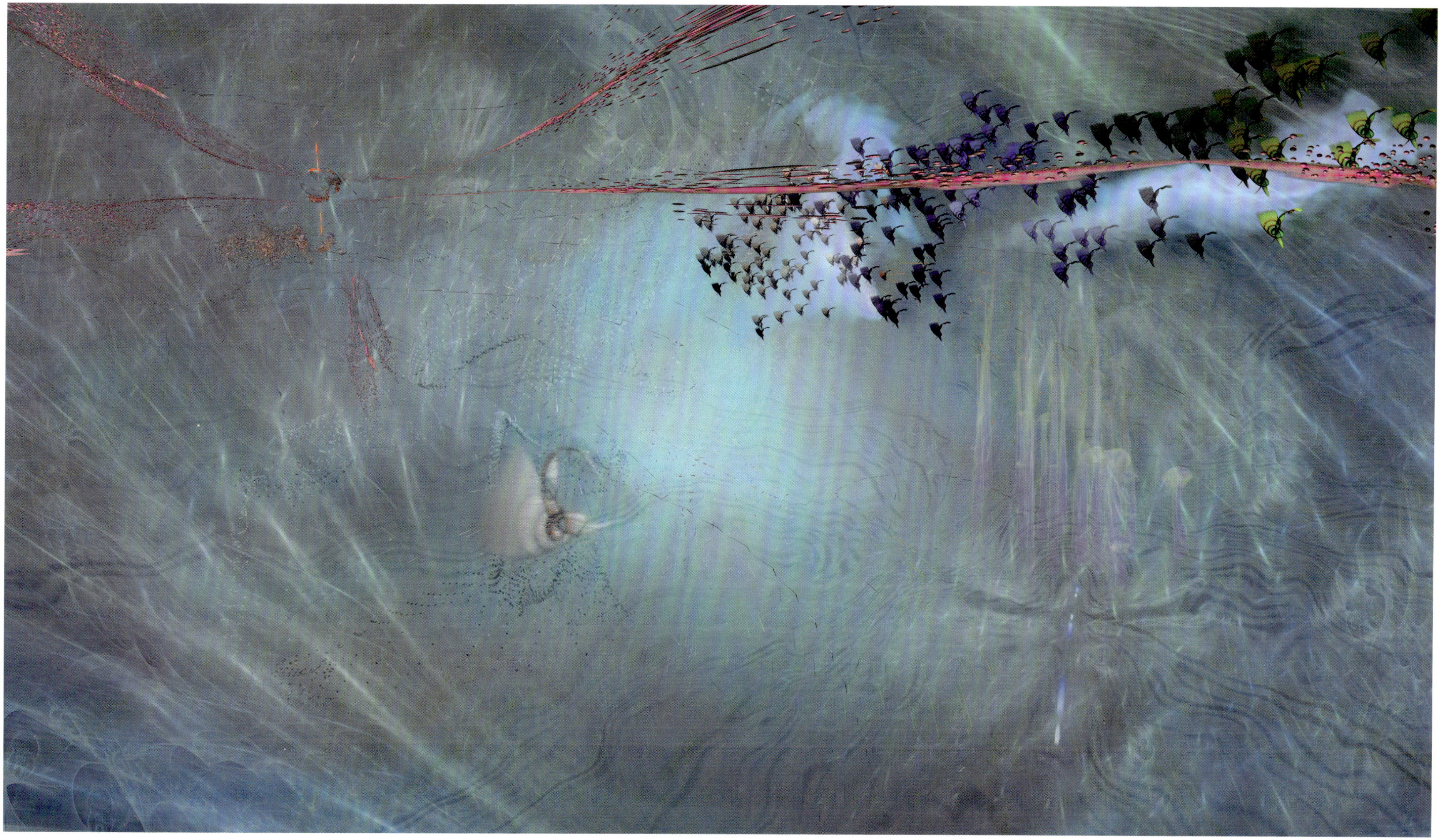

International Building, Rockefeller Center

Seagram Building

Guggenheim Museum

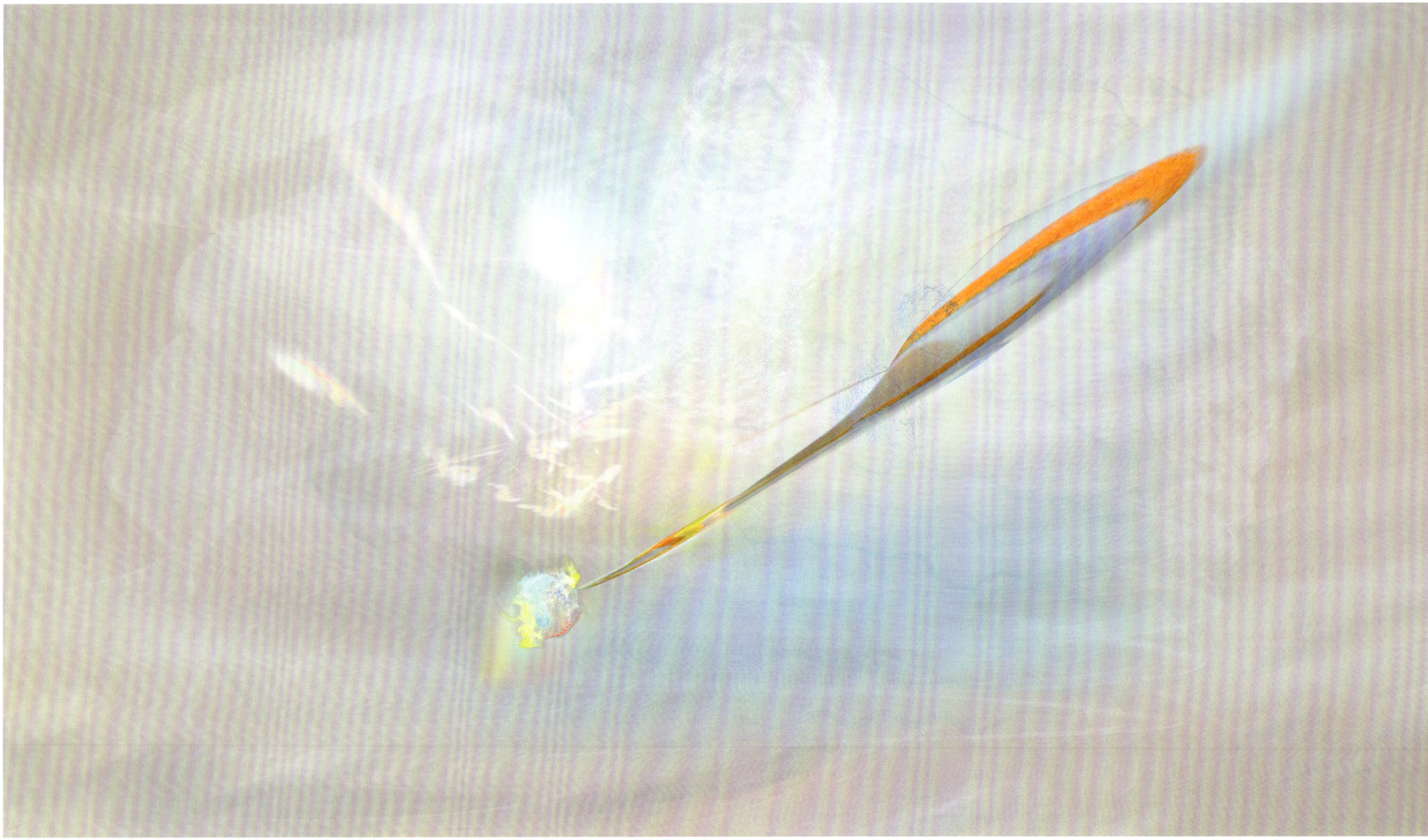

Central Park

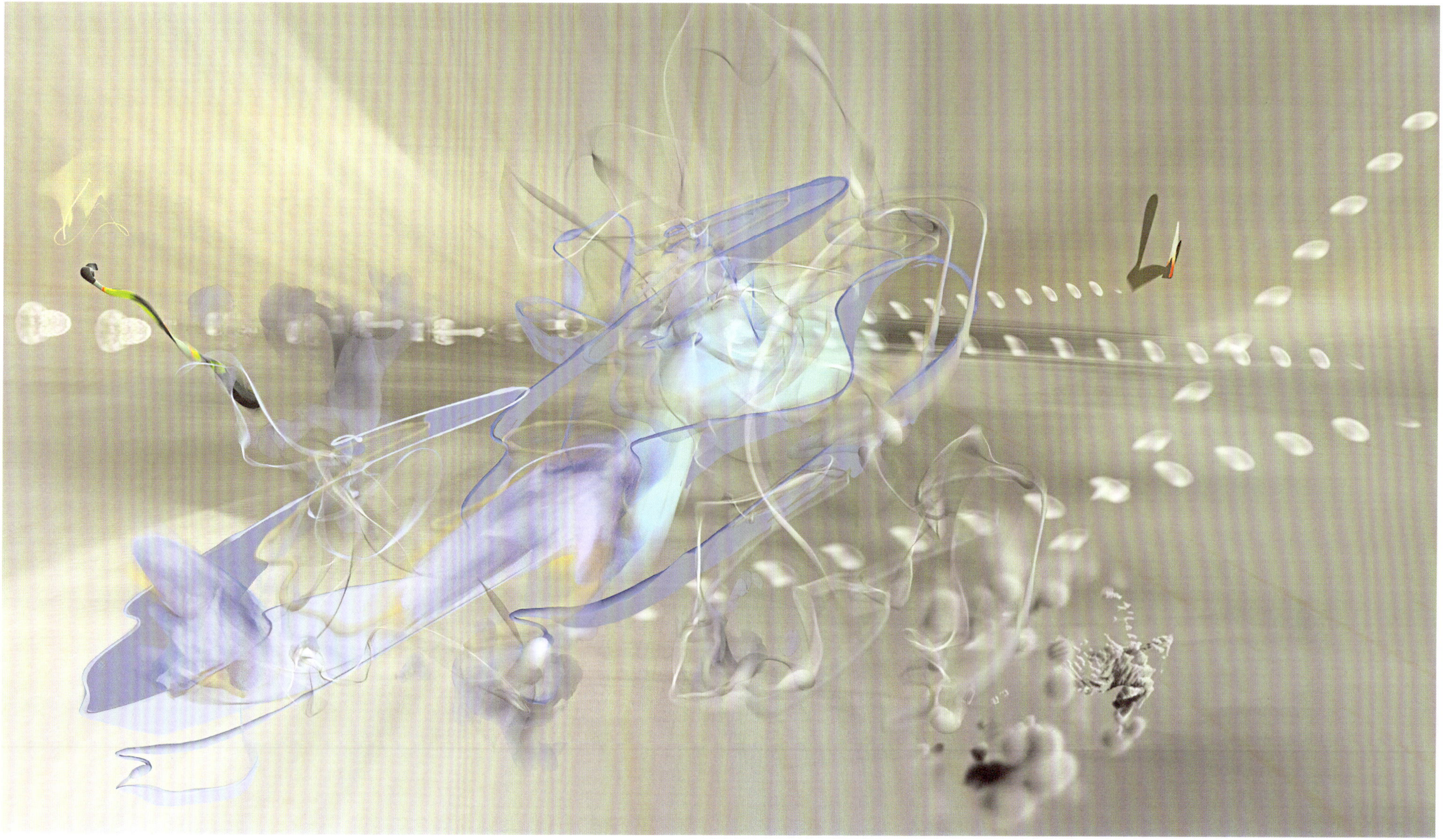

Academical Village

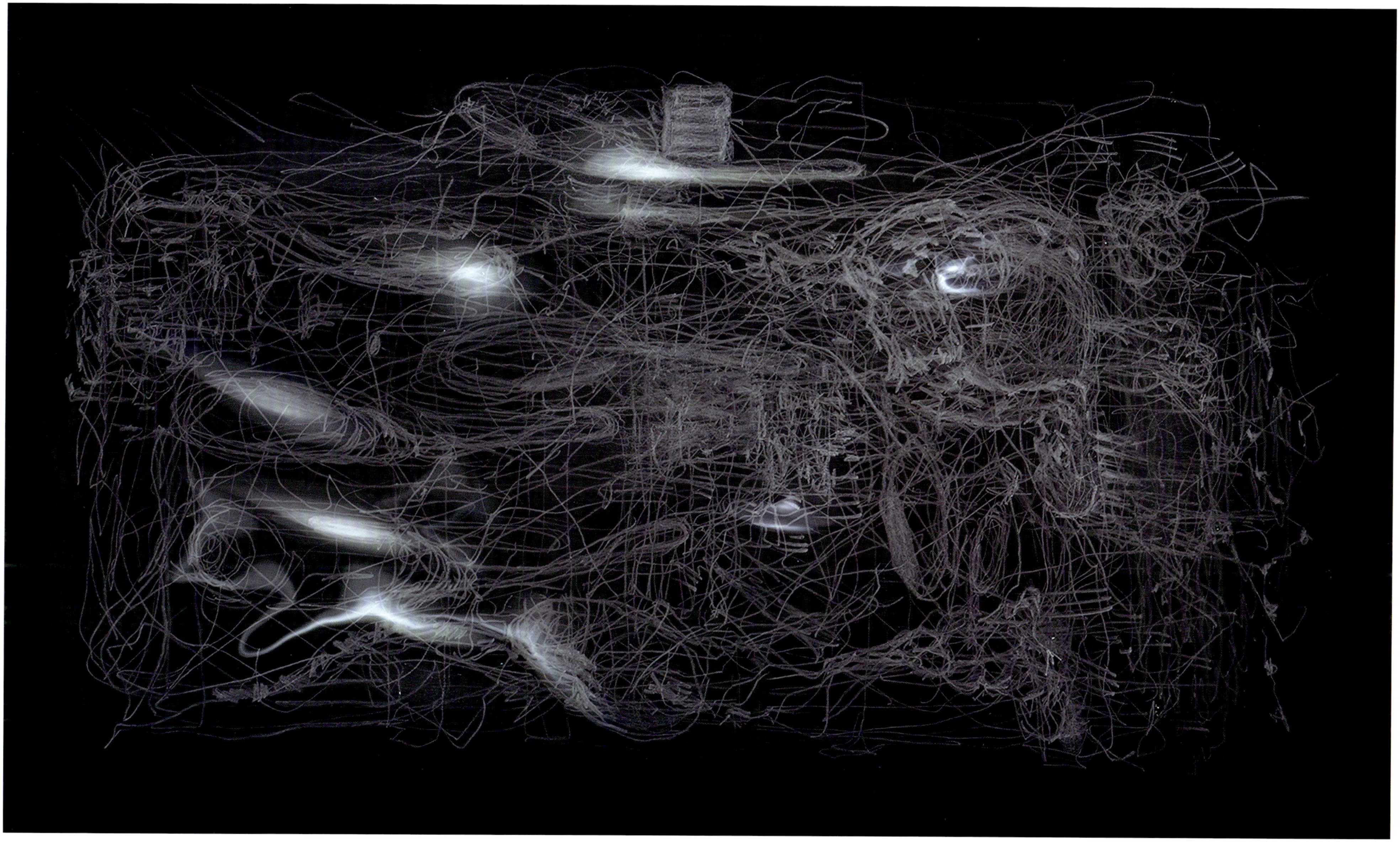

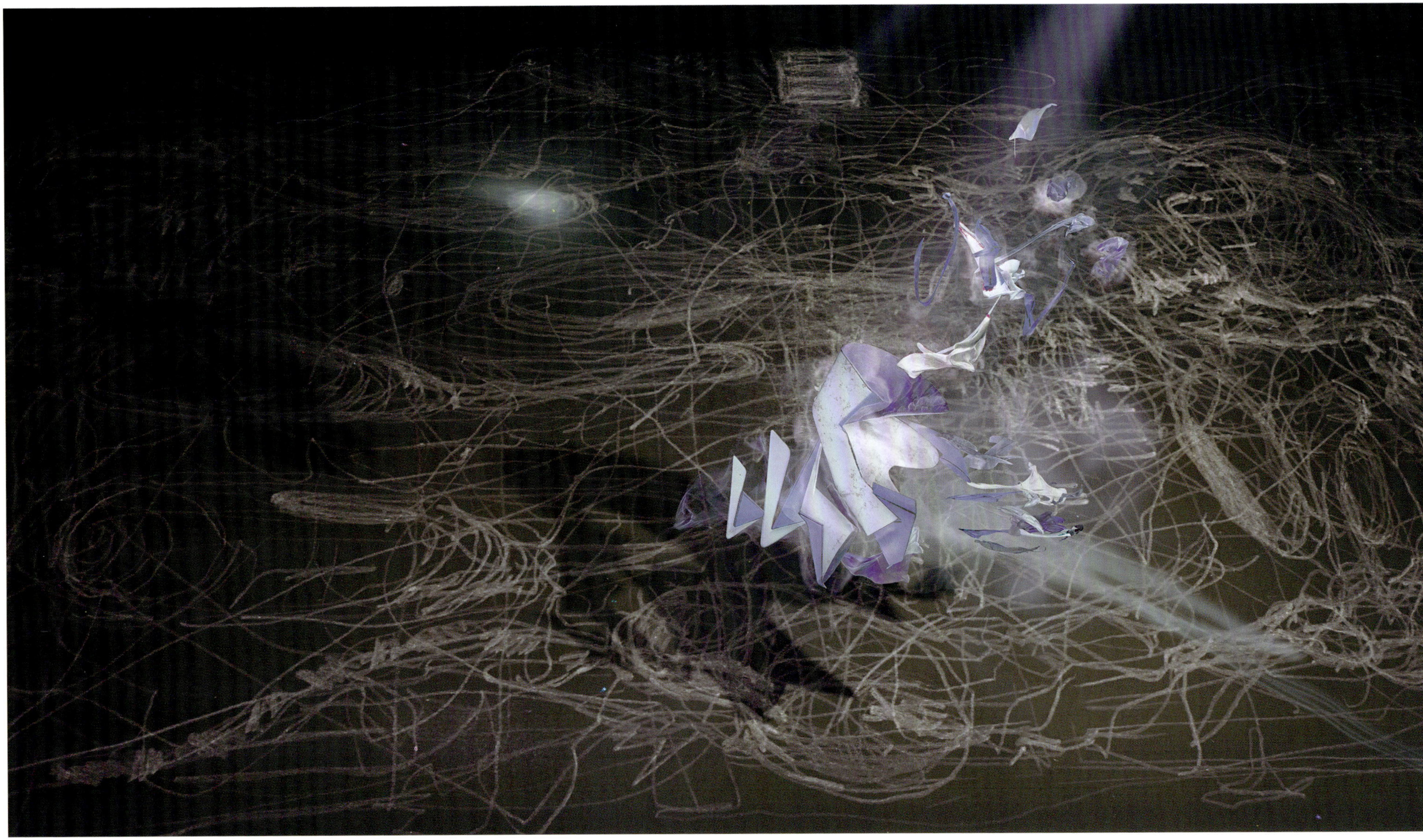

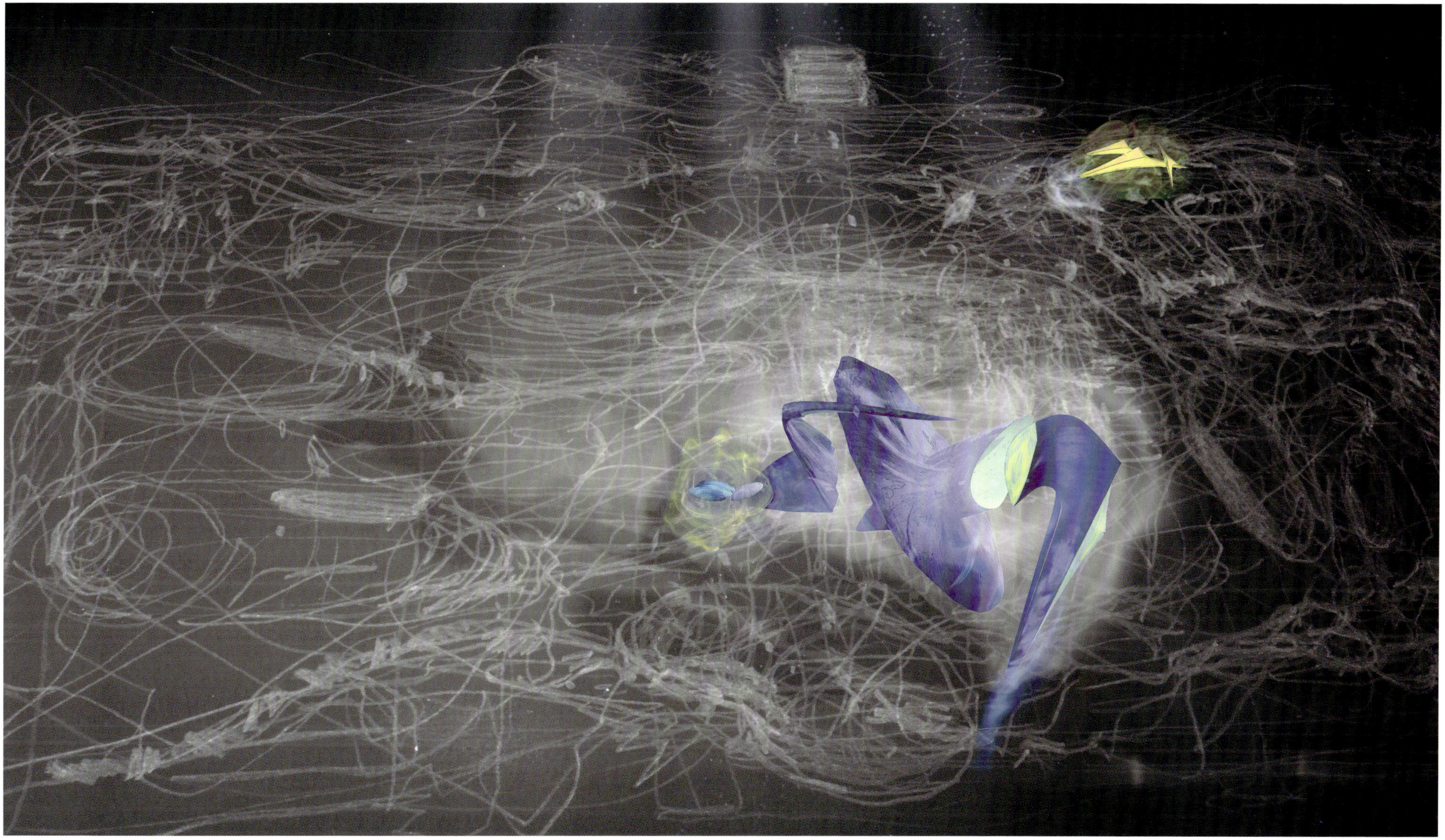

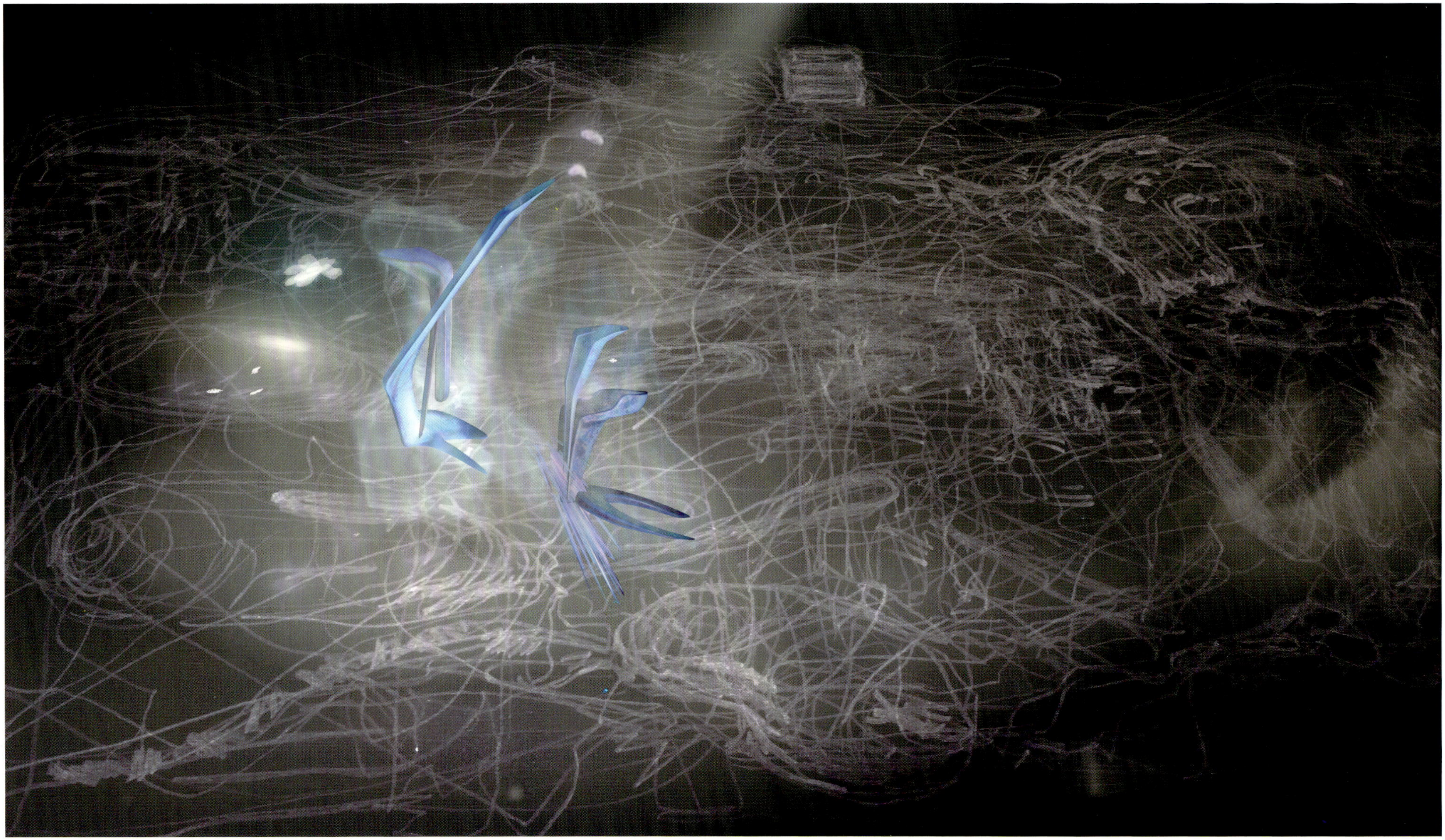

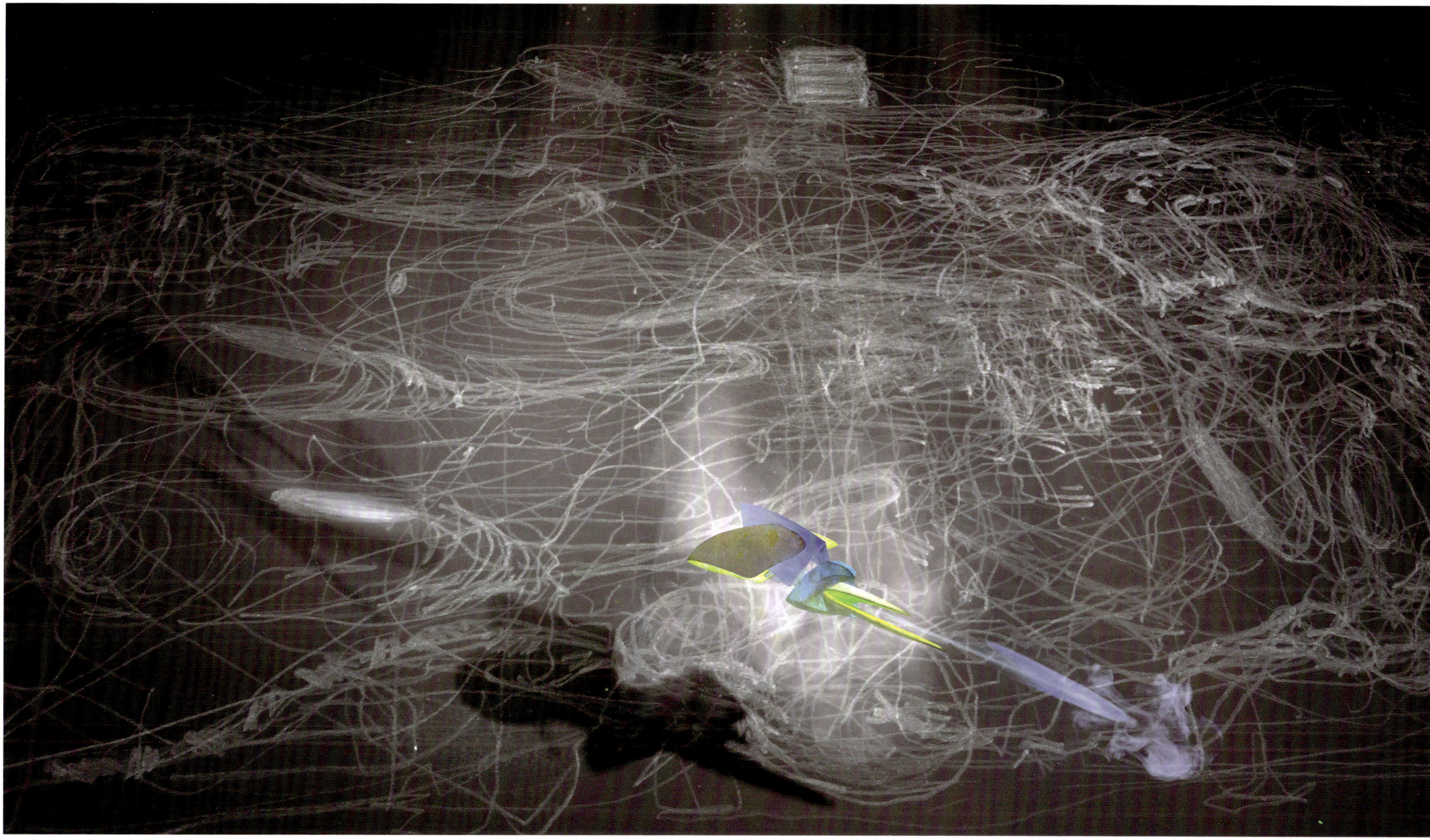

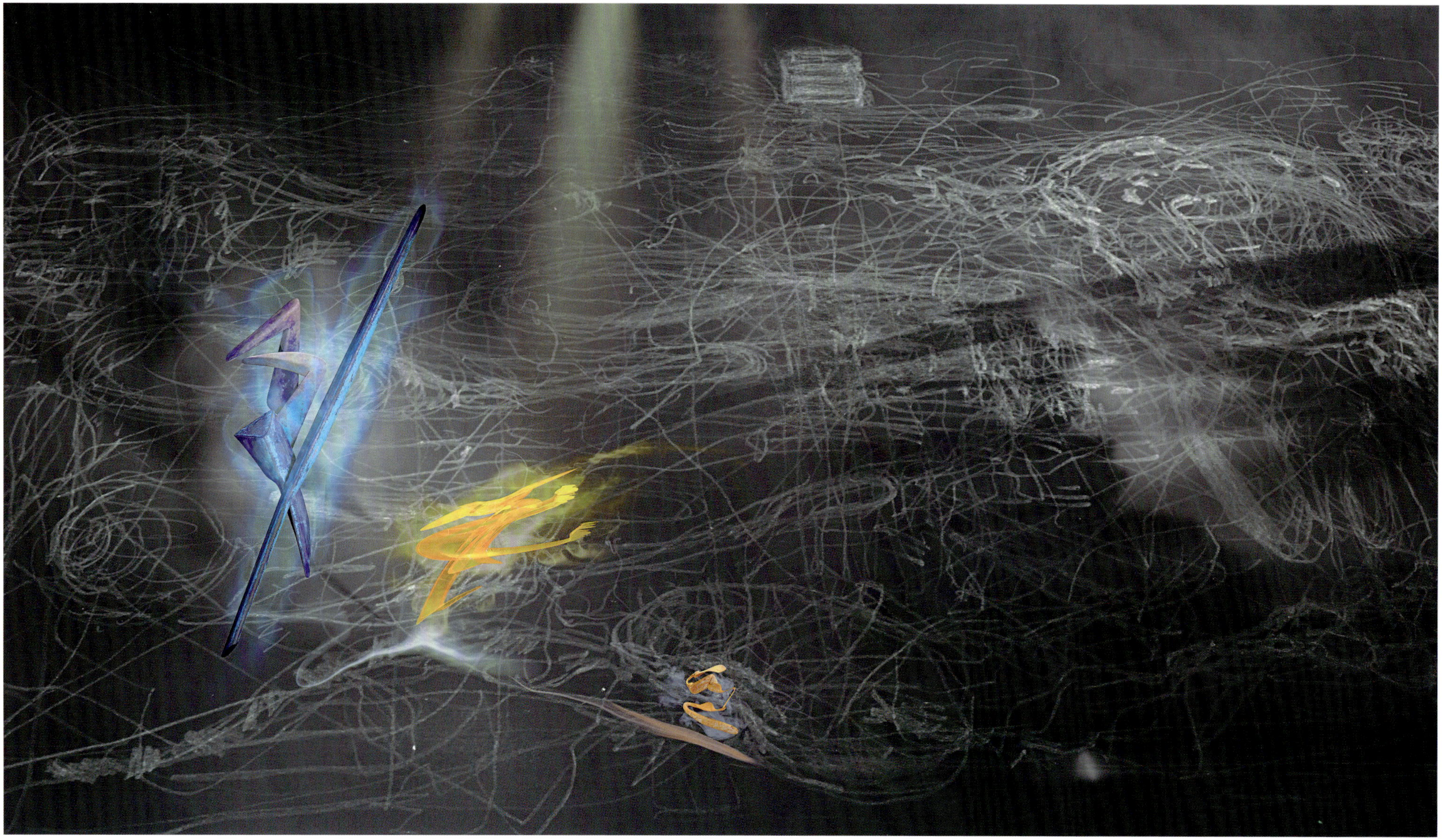

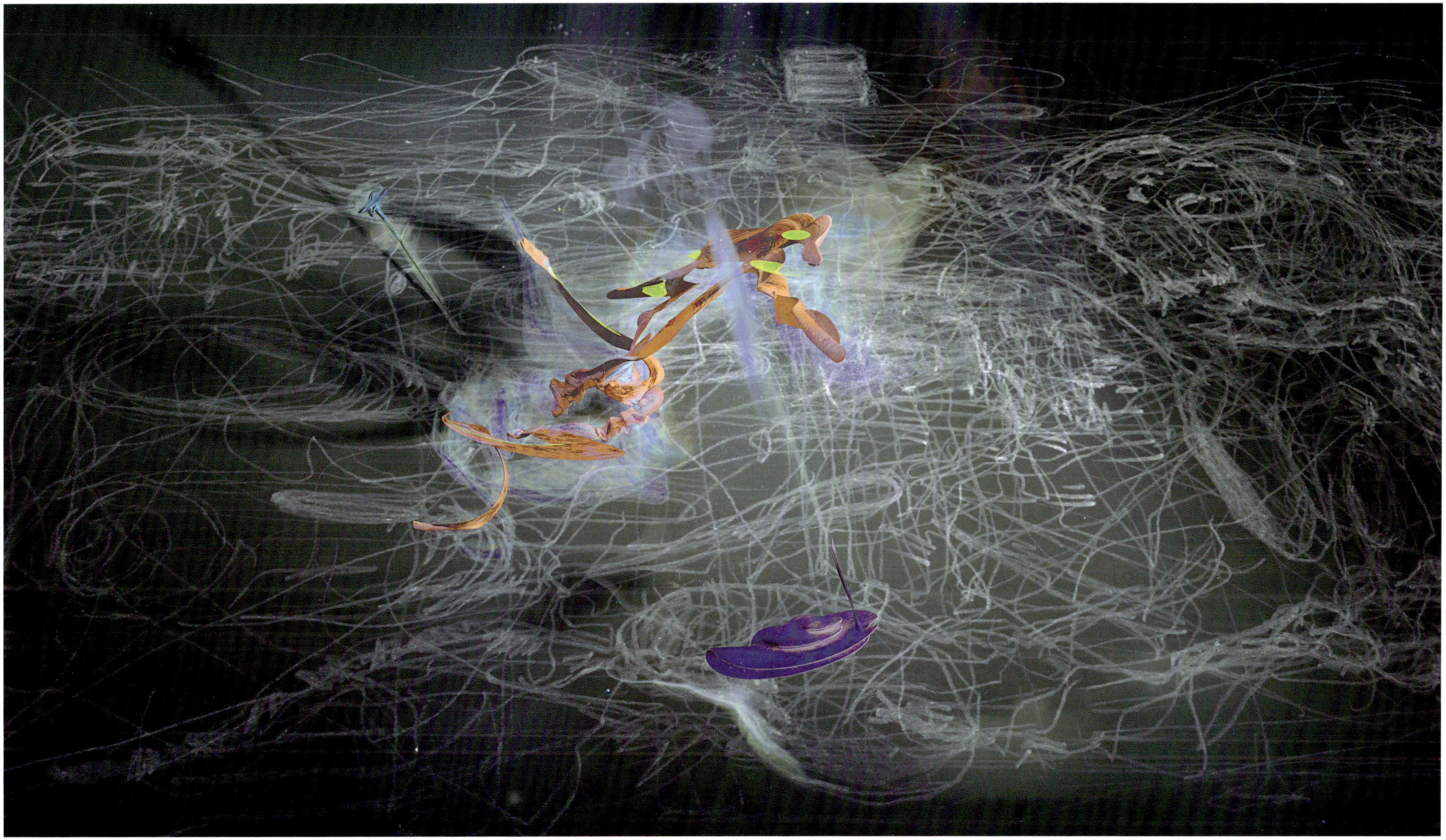

Notes

Preface

1 Böhme, Gernot. *The Aesthetics of Atmospheres*. Edited by Jean-Paul Thibaud. Routledge, 2017.

2 Böhme, *The Aesthetics of Atmospheres*, 2.

Introduction

This essay was adapted from an earlier chapter, "Architecture" in *The Handbook of Listening*, edited by Debra L. Worthington and Graham D. Bodie, 71–88. John Wiley & Sons, 2020.

1 Voegelin, Salomé. *Sonic Possible Worlds: Hearing the Continuum of Sound*. Bloomsbury Press, 2014, 3.

2 Bull, Michael. "Auditory." In *Sensorium: Embodied Experience, Technology, and Contemporary Art*, edited by Caroline. A. Jones, 112–114. MIT Press, 2006, 112.

3 Böhme, Gernot. *The Aesthetics of Atmospheres*. Edited by Jean-Paul Thibaud. Routledge, 2017, 2.

4 Hobart, Michael E., and Zachary S. Schiffman. *Information Ages: Literacy, Numeracy, and the Computer Revolution*. Johns Hopkins University Press, 1998, 14–15.

5 Waller, Steven J. "Auditory Illusions in the Soundscapes of Rock Art and Stonehenge." In *Archaeoacoustics: The Archaeology of Sound*, edited by Linda C. Eneix. OTS Foundation, 2014.

6 UNESCO. "Ħal Saflieni Hypogeum: Description." Accessed January 8, 2020. http://whc.unesco.org/en/list/130/.

7 Cook, Ian, Sarah K. Pajot, and Andrew F. Leuchter. "Ancient Architectural Acoustic Resonance Patterns and Regional Brain Activity." *Time and Mind* 1, no. 1 (2008): 95–104. https://doi.org/10.2752/175169608783489099.

8 Lubman, David. "Convolution-Scattering Model for Staircase Echoes at the Temple of Kukulkan." *Journal of the Acoustical Society of America* 123, no. 5 (2008): 4161–4166. https://doi.org/10.1121/1.2934779.

9 Ober, Josiah, and Barry Strauss. "Drama, Political Rhetoric, and the Discourse of Athenian Democracy." In *Nothing to Do with Dionysos? Athenian Drama in its Social Context*, edited by John J. Winkler and Froma I. Zeitlin, 237–270. Princeton University Press, 1990.

10 Sheridan, Ted, and Karen Van Lengen. "Hearing Architecture: Exploring and Designing the Aural Environment." *Journal of Architectural Education* 57, no. 2 (2003): 38.

11 Lawler, Robert. "Geometry at the Service of Prayer: Reflections on Cistercian Mystic Architecture." *Parabola* 3, no. 1 (1978): 12–13.

12 Sheridan and Van Lengen. "Hearing Architecture," 40.

13 Wittkower, Rudolf. *Architectural Principles in the Age of Humanism*. Academy Editions, 1998, 19.

14 Howard, Deborah, and Laura Moretti. *Sound and Space in Renaissance Venice: Architecture, Music, Acoustics*. Yale University Press, 2009.

15 Russolo, Luigi. "The Art of Noises." Translated from the Italian. Pendragon Press, 1986, 30. https://monoskop.org/images/0/09/Russolo_Luigi_The_Art_of_Noises.pdf

16 Russolo. "The Art of Noises," 75.

17 Düchting, Hajo. *Paul Klee: Painting Music*. Prestel Verlag, 2002, 10.

18 Thompson, Emily. *The Soundscape of Modernity: Architectural Acoustics and the Culture of Listening in America*, 1900–1933. MIT Press, 2002, 227.

19 Irwin, Robert. *Being and Circumstance: Notes Toward a Conditional Art*. Lapis Press, 1985, 28.

20 Diller Scofidio + Renfro. "Blur Building." Accessed November 1, 2022. https://dsrny.com/project/blur-building.

Presence

This essay was first published as Van Lengen, Karen, "Lacrimae Rerum and Tears of Memory: Sonic Encounters in the Public Realm" in *The Sound of Architecture: Acoustic Atmospheres in Place*, edited by Angeliki Sioli and Elisavet Kiourtsoglou, 203–14. Leuven University Press, 2022.

1 Böhme, Gernot. *The Aesthetics of Atmospheres*. Edited by Jean-Paul Thibaud. Routledge, 2017, 174.

2 Voegelin, Salomé. *The Political Possibility of Sound: Fragments of Listening*. Bloomsbury Academic, 2019, 47.

3 Lacey, Kate. *Listening Publics: The Politics and Experience of Listening in the Media Age*. Polity Press, 2013, 91.

4 AP News. "Remembering D-Day: Key facts and figures about the invasion that changed the course of World War II." https://apnews.com/article/dday-wwii-france-invasion-military-b02d03fa11f66767a521a3b01357a89a#. Accessed February 19, 2020.

5 Centre des Monuments Nationaux. "History of the Arc de Triomphe." https://www.paris-arc-de-triomphe.fr/en/discover/history-of-the-arc-de-triomphe. Accessed November 15, 2024.

6 Fontana, Bill. "Sound Island." *Resoundings*. https://www.resoundings.info/sound-island-paris-1994. Accessed August 27, 2020.

7 Young, James E. "Memory and Counter-Memory." *Harvard Design Magazine*, no. 9 (Fall 1999): unpaginated.

8 Knitz, Andreas, and Horst Hoheisel. "Aschrottfountain (Kassel, 1985)." http://www.knitz.net/index.php?option=com_content&task=view&id=30&Itemid=32&lang=en. Accessed March 7, 2020.

9 Young, "Memory and Counter Memory."

10 Knitz and Hoheisel. "Aschrottfountain."

An initial, longer version of this essay was published in von Fischer, Sabine, and Olga Touloumi, eds. "Sound Modernities: Histories of Media and Modern Architecture." Special issue, *Journal of Architecture* 23, no. 6 (2018): 1003–1029.

Communication

1 Keller, Helen. *Helen Keller in Scotland: A Personal Record Written by Herself*. Edited by James K. Love. Meuthen, 1933, 68. https://archive.org/details/helenkellerinscooohele/page/n9

2 Certeau, Michel. *The Practice of Everyday Life*. University of California Press, 1984, 158.

3 For more on Aino Marsio-Aalto see also: Stritzler-Levine, Nina. "Introduction." In *Artek and the Aaltos: Creating a Modern World*, edited by Stritzler-Levine, Nina, and Timo Riekko, 21–39. Bard Graduate Center, 2016; and Mikonranta, Kaarina. "Aino Marsio-Aalto—Interior and Furniture Designer." In *Aino Aalto*, edited by Ulla Kinnunen, 110–163. Alvar Aalto Museum, 2004; and Suominen-Kokkonen, Renja, ed. *Aino Aalto: A Shared Journey: Interpretations of an Everyday Modernism*, 17–40. Ram Publications, 2007.

4 Pelkonen, Eeva-Liisa. *Alvar Aalto: Architecture, Modernity, and Geopolitics*. Yale University Press, 2009, 1–10; Goldhagen, Sarah Williams. "Alvar Aalto's Astonishing Rationalism." In *A Field Guide to a New Meta-Field: Bridging the Humanities-Neurosciences Divide*, edited by Barbara Maria Stafford, 267–308. University of Chicago Press, 2011; Abercrombie, Stanley. "Woodberry Poetry Room." *Interiors*, February 1979: 74–75; Jormakka, Kari. "Poetry in Motion: Aalto's Woodberry Poetry Room at Harvard." In *Aalto and America*, edited by Stanford Anderson, Gail Fenske, and David Fixler, 251–262. Yale University Press, 2012; Mattern, Shannon. "Fluid Text, Total Design: The Woodberry Poetry Room as Idea, Collection, and Place." *Space and Culture* 14, no. 1 (2011): 27–50.

5 Metcalf, Keyes D. "The Architecture of the Lamont Library." UAIII 50.8.11.3, Box 58. Harvard University Archives (HUA).

6 "Harvard Faces Its Library Problems." In *The Place of the Library in a University: A Conference Held at Harvard University, 30–31 March, 1949*. Harvard University Library, 1950, 3–14. HUF525.531.26a, HUA.

7 Wecter, Dixon. "General Reading in a University Library." *Harvard University Bulletin*, 1950, 8.

8 Ingham, Travis. "Harvard Poetry Room, Housing Rare Works, Barred to Faculty." *Boston Herald*, June 12, 1931.

9 Frost, Robert. Robert Frost to John Bartlett, December 8, 1913. In *The Letters of Robert Frost, Volume I: 1886–1920*. Edited by Donald Sheehy, Mark Richardson, and Robert Faggen. Harvard University Press, 2014, 176.

10 "The University Film Foundation of Harvard University." *Science* 71, no. 1841 (April 11, 1930); and *Harvard Crimson*. "University Film Foundation to be Run by Harvard." April 24, 1934.

11 Eliot, T. S., in Sweeney, John L. "A Place for Poetry: The Woodberry Poetry Room in Widener and Lamont." *Harvard Library Bulletin* 8, no. 1 (1954): 71.

12 Furr, Derek. *Recorded Poetry and Poetic Reception from Edna Millay to the Circle of Robert Lowell*. Palgrave Macmillan, 2010, 5.

13 For Packard's proposal see: "Poetry Recordings (1938–1939): A Proposal for a Permanent Vocarium of Poetry Records in the Harvard College Library." UAIII 50.8.11.3, Box 82, HUA.

14 Sterne argues that the headset culture materialized a "collectivization of individuated listening." See: Sterne, Jonathan. *The Audible Past: Cultural Origins of Sound Reproduction*. Duke University Press, 2003, 154–77.

15 This is an argument most notably made by Demetri Porphyrios in *Sources of Modern Eclecticism: Studies on Alvar Aalto*. St. Martin's Press, 1982, 24.

16 Aalto, Alvar. "The Humanizing of Architecture: Functionalism Must Take the Human Point of View to Achieve Effectiveness." *Technology Review*, November 1940, 14–15.

17 Aalto, Alvar, in Pallasmaa, Juhani. "Alvar Aalto: Toward a Synthetic Functionalism." In *Alvar Aalto: Between Humanism and Materialism*, edited by Peter Reed (Museum of Modern Art, 1998), 30.

18 Hammarstrom, Olav. Letter from Olav Hammarstrom to Alvar Aalto, September 7, 1948. Poetry Room Folder, Alvar Aalto Museum. All translations from Swedish are kindly provided by Jennifer Mack.

19 Schildt, Göran. *Alvar Aalto: The Decisive Years*. Rizzoli, 1986, 14–15.

20 Alvar Aalto discusses this new role of the designer in relationship to the Stockholm Exhibition I, 1930. See: Aalto, Alvar. "Democracy, planning, and responsibility." In *Alvar Aalto, 1898–1976*. Edited by Aarno Ruusuvuori. Museum of Finnish Architecture, 1978, 101.

21 Åhrén, Uno, et al. "Acceptera." In *Modern Swedish Design: Three Founding Texts*, edited by Lucy Creagh, Helena Kåberg, and Barbara Miller Lane, 140–347. Museum of Modern Art, 2008.

22 Quantrill, Malcolm. *Alvar Aalto: A Critical Study*. New Amsterdam, 1989, 66; and Pelkonen, *Alvar Aalto*, 66–69.

23 Spens, Michael, and Alvar Aalto. *Viipuri Library, 1927–1935: Alvar Aalto*. Academy Editions, 1994, 35.

24 Aalto, Alvar, "Municipal Library in Viipuri: Description of the building's construction." In *Kaupunginkirjasto 1927, 1933–35 = Municipal Library, Viipuri*, edited by Päivi Lukkarinen. Alvar Aalto-Museo, 1997, 16.

25 Aalto, Alvar, in Pallasmaa. "Alvar Aalto: Toward a Synthetic Functionalism," 30.

26 Giedion, Siegfried. *Space, Time, and Architecture: The Growth of a New Tradition*. 5th ed. Harvard University Press, 1967, 633.

27 Frampton, Kenneth. *Modern Architecture: A Critical History*. 4th ed. Thames & Hudson, 2007, 198.

28 Huxtable, Ada Louise. "An Enduring Legacy." In *On Architecture: Collected Reflections on a Century of Change*. Walker, 2008, 186.

29 Aalto, Alvar. "The Humanizing of Architecture," 36.

30 Aalto, Alvar. "The Influence of Construction and Material on Modern Architecture." In *Synopsis: Painting, Architecture, Sculpture. Malerei, Architektur, Skulptur. Peinture, Architecture, Sculpture*. Birkhauser Verlag, 1970, 13.

31 Aalto, Alvar, in Pallasmaa, Juhani. "Tuberculosis Sanatorium, Paimio, 1929–1933." In *Artek and the Aaltos*, Stritzler-Levine and Riekko, 203.

32 I am indebted to Shannon Mattern's article for her poignant observation on the importance of Aalto's design gesture to encase the turntables. Mattern, "Fluid Text, Total Design," 32.

33 Thompson, Emily Ann. *The Soundscape of Modernity: Architectural Acoustics and the Culture of Listening in America, 1900–1933*. MIT Press, 2002, 169–228.

Discovery

1 Cage, John. *Silence: Lectures and Writings*. Wesleyan University Press, 2010, 17.

2 Artsy. "Gutai." Accessed November 12, 2024. https://www.artsy.net/gene/gutai.

3 Jirō, Yoshihara. "Gutai: Splendid Playground," exhibition panel, Guggenheim Museum, 2013.

4 Castrigno, Marissa. "The Competition: 33 Plans for Central Park in 1858." *Central Park Conservancy*, April 18, 2019. Accessed on October 15, 2024. https://www.centralparknyc.org/articles/plans-for-central-park.

5 Soundscape Architecture. "Central Park." Accessed November 1, 2021. http://soundscape.iath.virginia.edu/image-map/1347.

6 Soundscape Architecture. "Listening to the Lawn." Accessed November 1, 2018. http://soundscape.iath.virginia.edu/node/919#overlay-context=node/919.

7 Bachelard, Gaston. *Water and Dreams: An Essay on the Imagination of Water*. Translated by Edith R. Farrell. Pegasus Foundation, 2006, 22–23.

8 "New York Five" is a reference to the group of New York-based architects whose work was presented in the book *Five Architects*. Wittenborn, 1972. The architects included Peter Eisenman, Charles Gwathmey, Michael Graves, John Hejduk, and Richard Meier.

9 Rykwert, Joseph. *The Dancing Column: On Order in Architecture*. MIT Press, 1996.

10 Kwinter, Sanford. *African Genesis*. Recalled from his lecture at the University of Virginia, 2004.

Art

1 John Cage quoting Oskar Fischinger in an interview with Joel Eric Suben, 1983. In *Conversing with Cage*, edited by Richard Kostelanetz. Routledge, 2003, 8.

Acknowledgments

The work represented in this book has been made possible by the many people who have supported these ideas at the intersection of sound, architecture, and art. Among those who have inspired us in the development of our projects are Bill Fontana, Jean Gardner, David Hykes, and Ted Sheridan. We also thank our collaborators on *MIX House* and *Sound Lounge*, Joel Sanders and Ben Rubin.

We are deeply grateful to the contributing authors and commentators included in this book: Barry Bergdoll, Tony Hiss, Nana Last, Olga Touloumi, and Peter Waldman. In addition, we are indebted to the editors of two publications that included earlier versions of two of the essays found in this book: Angeliki Sioli and Elisavet Kiourtsoglou, editors of *The Sound of Architecture: Acoustic Atmospheres in Place*, and Debra L. Worthington and Graham D. Bodie, editors of *The Handbook of Listening*. We also wish to thank Elisabeth Fitzgerald for her editorial comments on our text.

At the School of Architecture at the University of Virginia, we thank Dean Malo Hutson and Emeritus Professor Reuben Rainey for their support and encouragement of this book. At the broader University, we are grateful for the excellent guidance we received from Worthy Martin and Lauren Massari at the Institute for Advanced Technology in the Humanities (IATH), and for the grant support of the Jefferson Trust. We also thank Leigh Wilkerson for her early graphic advice on the *Soundscape Architecture* website at IATH.

We thank our musical collaborators, Troy Rogers for his additional sound compositions of the New York City Soundscapes, and Matt Wyatt for his inspired musical composition of Central Park, using our recordings as primary material along with his own musical additions.

Over the years, we have had many students who have contributed to various phases of these projects. We thank Roderick Cruz, Nate Dansie, Alan Ford, Christina Griggs, Harsh Jain, Ryan Lewandowski, and Kaitlynn Long for their assistance.

We are grateful to ORO Editions, and to our publisher, Gordon Goff, project manager, Jake Anderson, and graphic design consultant, Pablo Mandel/Circular Studio for their guidance and enthusiasm to publish this work.

And finally, we thank our daughter, Kiri Van Lengen-Welty, who has not only supported the unusual nature of this artistic journey, but who has also served as both an excellent editor and graphic designer of this book.

Contributors

Barry Bergdoll is the Meyer Schapiro Professor of Art History at Columbia University. Professor Bergdoll's broad interests center on modern architectural history, with a particular emphasis on France and Germany since 1750. Trained in art history rather than architecture, he has an approach most closely allied with cultural history and the history and sociology of professions. He served as Philip Johnson Chief Curator of Architecture and Design at the Museum of Modern Art from 2007 to 2013. Bergdoll's interests also include the intersections of architecture and new technologies of representation in the modern period, especially photography and film.

Tony Hiss is the author of 15 books, including *The Experience of Place: A New Way of Looking at and Dealing with our Radically Changing Cities and Countryside*, *The View from Alger's Window: A Son's Memoir*, and *In Motion: The Experience of Travel*. Hiss was also a staff writer for *The New Yorker* for more than 30 years, and a visiting scholar at New York University. The National Recreation and Park Association's National Literary Award has praised his lifetime of "spellbinding and poignant writing." *Rescuing the Planet: Protecting Half the Land to Heal the Earth* is his 15th book, 10 years in the making.

Nana Last is an art and architecture theorist and critic. She is professor of architecture at the University of Virginia, where she founded the interdisciplinary PhD Program in the Constructed Environment. Her writing examines relations between architecture, art, philosophy, and science in modern and contemporary society. Her books include *Wittgenstein's House: Language, Space, and Architecture* and *Archive Matrix Assembly: The Photography of Thomas Struth 1978–2018*. She is currently working on a book that examines the epistemological impacts of the advent of big data on architecture, art, and urbanism. She received a PhD in Architecture and Art: History, Theory, and Criticism from the Massachusetts Institute of Technology and a master's degree in architecture from Harvard University.

Olga Touloumi is associate professor of architectural history at Bard College. Her research concerns spatial politics, architecture, and media during the 20th century. Her book *Assembly by Design: The United Nations and its Global Interior* examines the emergence of a new type of public space, "the global interior," situating it within mid-20th-century debates on media democracies and liberal internationalism. She is currently working on a microhistory of the architectural worker and crocheter Christine Benglia-Bevington, presenting a counternarrative to the history of modern architecture in the United States that sheds light on antiheroic positions and informal archives. Touloumi has coedited "Sound Modernities: Architecture, Media, and Design," a special issue of *The Journal of Architecture*, which investigates how acoustics and sound technologies transformed modern architectural culture during the 20th century; and *Computer Architectures: Constructing the Common Ground*, a volume about the exchanges between designers and computational technologists in Europe and North America.

Karen Van Lengen, a Fellow of the American Institute of Architects, is the William R. Kenan Professor of Architecture and former dean (1999–2009) of the School of Architecture at the University of Virginia. Her current research explores the role of sound in defining atmospheres of spaces. Her design work has been widely exhibited and is now in the permanent collections of several major museums. In Van Lengen's academic leadership roles, she chaired the Department of Architecture at Parsons School of Design (1995–9), where she founded the Design Workshop Program; and as dean at the University of Virginia, she developed interdisciplinary programs focused on the relation between ethics and aesthetics and introduced courses on the sonics of space. Van Lengen has authored several books, and numerous articles on themes related to the sonics of spaces, architecture, and landscape architecture. She began her professional career as an associate of I. M. Pei & Partners, before founding her own award-winning firm in New York City. She holds a BA from Vassar College and a MArch from Columbia University.

Peter Waldman is rumored to have quarried mica ever since his early childhood explorations of the wilderness of New York City more than half a century ago. He studied architecture from 1961–9, first at Princeton University and later as a Peace Corps volunteer in Arequipa, Peru. He served his apprenticeship in the studios of Richard Meier, briefly, and more substantially with Michael Graves. Since the 1970s, he has been an architect and educator, teaching first at Princeton, briefly at the University of Cincinnati, then at Rice University, and currently at the University of Virginia, where he is now firmly grounded in the Piedmont Condition. Waldman's extensive residential practice has been concerned with the Climatic House constructed according to Specifications for Construction executed by Nomads, Surveyors, and Lunatics. His fables of the Gardener and the Engineer manifest his profound respect for the spirit and resources of the renewable American urban condition. He is the author of *Lessons From The Lawn: The Word Made Flesh; Dialogues Between Citizens and Strangers* and *Connective Tissues: Ten Essays by University of Virginia Kenan Fellows 2001–2016*, both published by ORO Editions.

Jim Welty is an artist, a sculptor of large metal and copper constructions, and an animator of ideas and sonic landscapes. His recent animation work has captured the lyrical quality of the sounds of architectural spaces and places. This work, with partner Karen Van Lengen, has been exhibited and collected by major national institutions. As master printer (1975–87) to Frank Stella, Welty developed several printing projects including the famed Moby Dick series. He also developed his own body of sculptural work that received high acclaim. His large format copper wall sculptures, represented by the John Davis Gallery in New York City, led to a significant retrospective exhibition at the University of Virginia's Fralin Art Museum. He is a graduate of Rochester Institute of Technology and the University of Oxford's Ruskin School of Art.